6G Connectivity-Systems, Technologies, and Applications
Digitalization of New Technologies, 6G and Evolution

RIVER PUBLISHERS SERIES IN COMMUNICATIONS AND NETWORKING

Series Editors

ABBAS JAMALIPOUR
The University of Sydney
Australia

MARINA RUGGIERI
University of Rome Tor Vergata
Italy

The "River Publishers Series in Communications and Networking" is a series of comprehensive academic and professional books which focus on communication and network systems. Topics range from the theory and use of systems involving all terminals, computers, and information processors to wired and wireless networks and network layouts, protocols, architectures, and implementations. Also covered are developments stemming from new market demands in systems, products, and technologies such as personal communications services, multimedia systems, enterprise networks, and optical communications.

The series includes research monographs, edited volumes, handbooks and textbooks, providing professionals, researchers, educators, and advanced students in the field with an invaluable insight into the latest research and developments.

Topics included in this series include:-

- Communication theory
- Multimedia systems
- Network architecture
- Optical communications
- Personal communication services
- Telecoms networks
- Wifi network protocols

For a list of other books in this series, visit www.riverpublishers.com

6G Connectivity-Systems, Technologies, and Applications
Digitalization of New Technologies, 6G and Evolution

Editors

Ramjee Prasad

CTIF Global Capsule (CGC), Aarhus University, Denmark

Dnyaneshwar Shriranglal Mantri

Sinhgad Institute of Technology, Lonavala, India

Sunil Kumar Pandey

Institute of Technology & Science, Mohan Nagar, Ghaziabad, India

Albena Dimitrova Mihovska

CTIF Global Capsule (CGC), Aarhus University, Denmark

NEW YORK AND LONDON

Published 2024 by River Publishers
River Publishers
Alsbjergvej 10, 9260 Gistrup, Denmark
www.riverpublishers.com

Distributed exclusively by Routledge
605 Third Avenue, New York, NY 10017, USA
4 Park Square, Milton Park, Abingdon, Oxon OX14 4RN

6G Connectivity-Systems, Technologies, and Applications / by Ramjee Prasad, Dnyaneshwar Shriranglal Mantri, Sunil Kumar Pandey, Albena Dimitrova Mihovska.

Routledge is an imprint of the Taylor & Francis Group, an informa business

ISBN 978-87-7022-835-0 (hardback)
ISBN 978-87-7004-184-3 (paperback)
ISBN 978-10-4012-096-5 (online)
ISBN 978-10-0351-592-0 (master ebook)

While every effort is made to provide dependable information, the publisher, authors, and editors cannot be held responsible for any errors or omissions.

Contents

Preface

यदा सत्त्वे प्रवृद्धे तु प्रलयं याति देहभृत् ।
तदोत्तमविदां लोकान् अमलान्प्रतिपद्यते ॥ १४-१४ ॥

Transliteration

yadā sattvē pravṛddhē tu pralayaṃ yāti dēhabhṛt |
tadōttamavidāṃ lōkān amalān pratipadyatē || 14-14 ||

Anvaya

Yada... When; deha-bhrt... the embodied being; yati... transitions;
pralayam... death; pravrddhe... predominated; sattve... by the mode of
goodness; tada... at the time; yati... reaches; amalan... the pure;
lakan... planets; uttama-vidam... with the highest knowledge;
pratipadyate... are attained.

Translation

"When the embodied being transitions death predominated by the mode of
goodness; at that time the pure planets with highest knowledge are attained"

The Bhagavad Gita (Chapter-14, Verse -14)

The evolution of wireless communication has been marked by significant
advancements and improvements in technology, leading to the development
of multiple generations of mobile networks, commonly referred to as 1G, 2G,
3G, 4G, and 5G. Each generation has introduced new features, increased data
speeds, improved network capacity, and enhanced user experiences.

The journey from 1G to 5G has been very transformative with every
generation starting from 1980s when the first commercially available wireless network with 1G launched with the usage of analog signals for voice

communication only and limited data capabilities and low voice quality. This was succeeded by the 2G (second generation) introduced in early 1990s and marked a significant leap in mobile communication technology. The 2G witnessed transition from analog to digital signals, enabling improved voice quality, higher capacity with introduction of GSM and CDMA, and supported basic data services like SMS (short message service).

The 3G (third generation), launched in early 2000s, brought higher data transfer rates, improved multimedia capabilities, enabled faster Internet browsing, video calling and mobile applications with introduction of technologies like UMTS (universal mobile telecommunications system) and HSPA (high-speed packet access) and supported mobile broadband services with data speeds typically ranging from 384 kbps to several Mbps. These developments laid down the foundation of a significant leap in mobile data speeds and network capacity with 4G (Fourth Generation). Deployed in the late 2000s, 4G offered much higher data transfer rates, providing a more seamless user experience for video streaming, online gaming, and other data-intensive applications with implementation of technologies like LTE (long-term evolution) and WiMAX (Worldwide Interoperability for Microwave Access) and supported IP-based voice and multimedia services (VoIP and video conferencing). All these developments kept the evolution of wireless communication going on and, commercially launched in the 2010s, the 5G which offered significantly higher data speeds, lower latency, and increased network capacity by utilizing advanced technologies like millimeter waves, massive MIMO (multiple-input multiple-output), and network slicing. The 5G also supports a wide range of applications, including IoT (Internet of Things), AR/VR, remote surgery, smart cities, and more.

The 5G revolutionized various industries and transformed the way we interact with technology, providing a foundation for emerging technologies and applications that require ultra-fast and reliable connectivity. As technology continues to advance, the wireless communication landscape will continue to evolve, paving the way for even more exciting innovations in the future with next-generation networks by applying intelligent networks capable of dealing with various network challenges with increased volume, density, complexities, and security with 6G. Several key drivers and expectations contribute to the push for 6G connectivity, are broadly include:

Data volume and speed demands: With the growing prevalence of data-intensive applications, such as ultra-high-definition video streaming, virtual and augmented reality, and cloud-based services, the volume of data being transmitted over mobile networks is exploding. 6G aims to offer even higher data speeds, potentially reaching multi-terabit per second rates, to meet the demands of such data-heavy applications.

Ultra-low latency: Some applications, like real-time gaming, autonomous vehicles, and remote robotic control, require ultra-low latency to function effectively and safely. 6G is expected to significantly reduce latency, possibly reaching sub-millisecond levels, enabling seamless and instantaneous interactions between devices and the network.

Massive IoT connectivity: The Internet of Things (IoT) is set to become even more pervasive, with billions of connected devices expected to be deployed across industries. 6G seeks to provide the capacity and efficiency to support this massive influx of IoT devices, allowing seamless connectivity and communication between devices and the cloud.

Enhanced spectral efficiency: To cope with the growing number of connected devices and the demand for higher data rates, 6G is expected to improve spectral efficiency. This means that 6G networks will be able to transmit more data using the same amount of spectrum, optimizing the utilization of the available frequency bands.

Enabling new applications and use cases: 6G is anticipated to enable novel applications and use cases that were not feasible with previous generations. This includes holographic communication, real-time translation, advanced AI-powered services, and immersive extended reality experiences, among others.

Integrated satellite and terrestrial networks: 6G is envisioned to integrate terrestrial and satellite networks, providing global coverage and connectivity in remote and underserved areas. This integration can expand the reach of communication networks and bridge the digital divide.

Quantum communication and enhanced security: Quantum communication is being explored as a potential security solution in 6G networks. Leveraging the principles of quantum mechanics, 6G could offer unprecedented levels of security, protecting against potential threats like quantum computing-based attacks.

Next-level connectivity for smart cities: 6G is expected to play a crucial role in the development of smart cities, supporting a myriad of applications like smart transportation, energy management, public safety, and environmental monitoring, among others.

Green and sustainable connectivity: With the increasing concern for environmental sustainability, 6G development is likely to focus on energy-efficient technologies and techniques to minimize the carbon footprint associated with network infrastructure.

The advent of 5G technologies has already started revolutionizing the way we connect and interact with the digital world. However, the tech industry never stands still, and research and development for the next generation of wireless communication, 6G, is already underway. This book explores the potential of 6G connectivity, usage in advanced technologies, and exciting applications that it promises to bring. From ultra-high data speeds and ultra-low latency to massive IoT support and revolutionary applications, 6G is poised to transform industries and reshape the digital landscape. As we take a glimpse into the future, 6G connectivity presents unparalleled opportunities, challenges, and promises for a hyper-connected world.

As the world anticipates the next leap in wireless communication, 6G connectivity emerges as a promising paradigm that goes beyond the capabilities of 5G technology. The book explores various aspects and dimensions of 6G systems, connectivity technology, and applications by providing an overview of the key differences between 5G and 6G and discusses the motivations driving the development of 6G. It then delves into the fundamental architectural components and novel technologies that are envisioned to empower 6G networks. Furthermore, the book also presents various real-world applications that could leverage the unprecedented speed, ultra-low latency, and enhanced reliability offered by 6G. Additionally, the challenges and potential solutions in the implementation of 6G connectivity are explored. By comprehensively examining the landscape of 6G, this book aims to offer valuable insights into the transformative potential of this technology in shaping the future of wireless communication.

In summary, the need for 6G connectivity arises from the pursuit of higher data speeds, ultra-low latency, massive IoT connectivity, enhanced spectral efficiency, and the facilitation of new and transformative applications. By addressing these drivers and expectations, 6G aims to revolutionize wireless communication, opening up a realm of possibilities for industries, societies, and individuals.

This book has 12 chapters.

Chapter 1: Intelligent Security for DDoS in HetIoT (6G Perspective): It presents the details about the DDoS attacks on the heterogeneous Internet of Things infrastructure using ARE approach. It also discusses the challenges, issues, and solutions.

Chapter 2: Industry 5.0 and 6G: Human-centric Approach: This chapter examines the essential requirements, and enabling technologies used in Industry 5.0 with 6G evolutions. It puts forward a proposal for enhancement of the 4.0 to 5.0 vision of Industrial Revolutions using next-generation networks and technologies such as AI, IoT, cloud computing, robotics, data analytics, etc., in order to achieve the goal. It also gives an idea about the COBOTS and human–machine interactions for increased productivity. The chapter is concluded with proposal and challenges in Industry 5.0.

Chapter 3: Role of 6G, IoT with Integration of AI and ML and Security in Agriculture: This chapter focuses on the fundamental concepts, architecture, and applications of 6G, IoT, AI, and ML, emphasizing their integration to establish a secure communication environment. It explores the empowering applications of IoT integrated with AI and machine learning using 6G communication networks.

Chapter 4: Visible Light Communications for 6G: Motivation, Configurations, and New Materials: It focuses on the optical communications along with its properties, challenges. However, it shows that taking into consideration the optical propagation, even with the proposed alternative coniïñAgurations, the MIMO capacity does not scale as the number of transmitters grows from some speciiïñAc density of optical APs in the area of interest.

Chapter 5: Access Security in 6G: The 6G-ACE Protocol (A Concept Proposal): This chapter analyzes history, the current situation, and what the access security needs of 6G might be. It also presents a concept proposal of what 6G subscriber access security protocol could be. The design of the "6G Authentication and Context Establishment" protocol (6G-ACE) will focus on principles and concepts, while avoiding detail when possible. The solution should be fast, effective, and reliable, and it should provide security contexts tailored to the task at hand. A 6G solution should break with existing solution, when necessary, but it should also seek to retain proven designs when feasible.

Chapter 6: ICT Applications in Health Monitoring: In this chapter, the directions of health research at Huawei for vital signs monitoring will be introduced. The convenience and reliability of medical technology applications can only be guaranteed with the appropriate implementation of efficient interfacing electronics and novel sensor solutions (e.g., MEMS, optical, micro-fluidics, etc.) especially in consumer products.

Chapter 7: Key Issues in NOMA from a 6G Perspective: This chapter focuses on exploring various key research issues that need to be tackled for the performance enhancement of NOMA regarding 6G.

Chapter 8: Green Computing: Importance, Approaches, and Practices:
This chapter aims to uncover the various dimensions including optimization
of energy consumption, approaches to reduce biological hazards (H/Gases,
etc.) and effective techniques to design environment friendly framework with
recycled uses of hardware and the applications of green computing. The chapter also emphasizes on latest techniques and approaches of green computing,
green design and developments, and approaches for energy optimization of
data centers. The chapter also discusses about present industry standards of
green computing, renewal green resources of energy.

**Chapter 9: Artificial Intelligence and Green 6G Network-enabled Archi-
tectures, Scenarios, and Applications for Autonomous Connected Vehicles:** In this chapter, we will examine the green 6G network-enabled architectures, scenarios, challenges, and Internet of Vehicles (IoV) applications for
future ACVs. We also discuss the applicability of AI and machine learning
(ML) paradigm for ACVs through several use cases.

**Chapter 10: Latest Advances on Deterministic Wired/Wireless Indus-
trial Networks:** This chapter investigates the most recent advances in deterministic OFDMA-based Wi-Fi for industrial applications. It concentrates on
techniques that provide deterministic support in industrial networks where
Wi-Fi is used as an additional technology to a deterministic wired core.

**Chapter 11: Cyber Threat Detection in 6G Wireless Networks using
Ensemble Majority-voting Classifier:** This chapter proposes an ensemble
majority-voting classifier for intrusion detection systems, to achieve the best
results based on the above-mentioned performance metrics. The accuracy of
the results for the proposed ensemble majority-voting classifier is 90.45%.

**Chapter 12: From Connectivity to Intelligence: Integrating IoT-6G for
the Future:** This chapter discusses regarding the enhancement of capacity of
6G networks by integrating it with IoT.

Ramjee Prasad
CTIF Global Capsule (CGC), Aarhus University, Denmark

List of Figures

List of Tables

List of Contributors

Al-Sakkaf, Ahmed Gaafar Ahmed, *Universidad Carlos III de Madrid, Spain*

Armada, Ana García, *Universidad Carlos III de Madrid (UC3M), Spain*

Arora, Varun, *Institute of Technology and Science, India*

Deshmukh, Madhukar, *Sharda University Greater Noida, India*

Dhar, Puja, *Institute of Technology and Science, India*

Javid, Iqra, *Sharda University Greater Noida, India*

Kansal, Smita, *Institute of Technology and Science, India*

Khara, Sibaram, *Sharda University Greater Noida, India*

Køien, Geir Myrdahl, *University of South-Eastern Norway (USN), Norway*

Kulkarni, Nandkumar Prabhakar, *MIT-ADT University, India*

Kumar, Rampravesh, *Birla Institute of Technology, India*

Kumar, Sanjay, *Birla Institute of Technology, India*

Mahadik, Shalaka Shankar, *Birla Institute of Technology Pilani, UAE*

Mantri, Dnyaneshwar Shriranglal, *Sinhgad Institute of Technology*

Mohan, Seshadri, *University of Arkansas at Little Rock, USA*

Morales-Céspedes, Máximo, *Universidad Carlos III de Madrid, Spain*

Muthalagu, Raja, *Birla Institute of Technology Pilani, UAE*

Pandey, Sunil Kumar, *Institute of Technology and Science, India*

Parthsarathy, Kavya, *Birla Institute of Technology Pilani, UAE*

Pawar, Pranav Motabhau, *Birla Institute of Technology Pilani, UAE*

Prasad, Neeli Rashmi, *TrustedMobi "VehicleAvatar Inc," USA*

Prasad, Ramjee, *Department of Business Development and Technology, Aarhus University, Denmark*

Raj, Tuhina, *Sharda University Greater Noida, India*

Ray, Abhay Kumar, *Institute of Technology and Science, India*

Saxena, Saurabh, *Institute of Technology and Science, India*

Sharma, Karan, *Birla Institute of Technology Pilani, UAE*

Sharma, Sachin, *Graphic Era Deemed to be University, India*

Singh, Kumar Pal, *Institute of Technology and Science, India*

Sofia, Rute C., *IIoT Competence Center, Fortiss, Germany*

Srivastava, Saurabh, *Birla Institute of Technology, India*

Tyagi, Ranu, *Graphic Era Deemed to be University, India*

Vaibhav Bhatnagar, Kumar, *Institute of Technology and Science, India*

Weigel, Walter, *Huawei European Research Institute, Belgium*

Wipiejewski, Torsten, *Huawei European Research Institute, Belgium*

Xu, Yaxin, *Huawei European Research Institute, Belgium*

List of Acronyms

3G	Third-generation
5G	Fifth-generation
6G	Sixth-generation
AC	Access categories
ACO-OFDM	Asymmetrically clipped optical OFDM
ACV	Autonomous connected vehicles
ADC	Analog-to-Digital Converter
ADC	Analog-to-digital converter
ADR	Angle diversity receiver
AGV	Automated guided vehicles
AI	Artificial intelligence
AIFS	Arbitration inter-frame space
AIFSN	Arbitration inter frame spacing number
AKA	Authentication and key agreement
AMR	Automated mobile robots
AP	Access Point
AP	Access point
ASME	Access security management entity
AUTN	Authentication token
AWGN	Additive white Gaussian noise
BAR	Block ack request
BCI	Brain—computer interface
BIA	Blind interference alignment
BIoT	Battle-field Internet of Things
BQRP	Bandwidth query report poll
BRP	Beamforming report poll
BS	Base station
BSR	Buffer status report
BSRP	Buffer status report poll
BSS	Basic service set
CAGR	Compound annual growth rate
CAV	Connected autonomous vehicles

CBS	Credit-based shaper
CCA	Clear channel assessment
CD-NOMA	Code domain NOMA
CFP	Contention free period
CI	Common information
CIAP	Confidentiality, integrity, availability, and privacy
CID	Context identifier
CIOT	Consumer Internet of Things
CNN	Convolutional neural network
COBOT	Collaborative robots
CoS	Class of Service
CP	Contention period
CPCS	Cyber physical cognitive systems
CR	Cognitive radio
CSI	Channel state information
CSMA/CA	Carrier sense multiple access with collision avoidance
CSMA/CD	Carrier sense multiple access with collision detection
CTS	Clear-to-send
CV	Cross-validation
CW	Contention window
D2D	Device to device
D2M	Device to machine
DBN	Deep belief network
DC	Direct current
DCF	Distributed coordination function
DCO-OFDM	Direct-current offset OFDM
DD	Direct detection
DDoS	Distributed denial of service
DDQN	Double deep Q-network
DFS	Design of the flexible distributed file system
DL	Deep learning
DL	Downlink
DQ	Q-Learning
DRL	Deep reinforcement learning
DSCP	Differentiated services code point
EC	Electricity conductivity
ECIES	Elliptic curve integrated encryption scheme

EDCA	Enhanced distributed channel access
EE	Energy efficiency
EPA	Environmental protection agency
FD	Full-duplex
FDD	Frequency division duplex
FL	Federated learning
FoV	Field of view
FoV	Field of view
FPA	Fixed power allocation
FTM	Fine timing measurement
FTPA	Fractional transmit power allocation
GCR MU-BAR	Group cast with retries multi-user block ack request
GMM	Gaussian mixture model
GPA	Generalized power allocation
HBC	Human bond communications
HC	Home security context
HC	Hybrid controller
HCCA	Hybrid coordination channel access
HCF	Hybrid coordination function
HE	Home environment
HetIoT	Heterogeneous Internet of Things
i.i.d	Independent and identically distributed
ICAIMRE	International Conference on Artificial Intelligence in Manufacturing and Renewable Energy
ICSs	Industrial control systems
ICT	Information communication technology
IDS	Intrusion detection systems
IFS	Interframe spacing
IIoT	Industrial Internet of Things
IM	Intensity modulation
IM	Intensity modulation
IMT	International Mobile Telecommunications
IoE	Internet of Everything
IoMT	Internet of Military Things
IoT	Internet of Things
IoV	Internet of Vehicles
iPCF	Industrial Point Coordination Function

IRS	Intelligent reconfigurable surfaces
IRS	Intelligent reconïñAgurable surfaces
IS	Intelligent security
IT	Information technology
ITU	International Telecommunication Union
KNN	K-nearest neighbor
KPI	Key performance indicator
KPI	Key performance indicator
LDR	Light dependent resistor
LED	Light emitting diodes
LED	Light emitting diodes
LLDP	Link layer discovery protocol
LLR	Log-likelihood rate
LoS	Line of sight
LoS	Line of Sight
LR	Logistic regression
LS	Latency sensitive
LSTM	Long short-term memory
LTE	Long-term evolution
M2M	Machine-to-Machine
MaaS	Mobility-as-a-Service
MAC	Media access control
MAP	Maximum a posteriori
MEMS	Microelectromechanical systems
MEMS	Micro-electro mechanical systems
MIMO	Multiple-input multiple-output
ML	Machine learning
MLME	MAC layer management entity
MM	Mapping matrix
MOTAG	Moving target defense
MPA	Message-passing algorithm
MPDU	MAC protocol data unit
MSTP	Multiple spanning tree protocol
MU-RTS	Multi-user request-to-send
NAV	Network allocation vector
NB-IoT	Narrowband IoT
NDP	Null data packet
NETCONF	Network configuration protocol
NFRP	Ndp frame report poll

NLoS	Non-line of sight
NLoS	Non-Line of Sight
NLP	Natural language processing
NMT	Nordic Mobile Telephony System
NOMA	Non-orthogonal multiple access
NPK	Nitrogen, phosphorus, and potassium
OFDMA	Orthogonal frequency division multiple access
OMA	Orthogonal multiple access
OP	Outage probability
OPA	Optimum power allocation
OpenCV	Open source computer vision
OWC	Optical wireless communication
OWC	Optical wireless communication
P2M	People to machine
P2P	People to people
PAM	Pulse amplitude modulation
PAM	Pulse amplitude modulation
PCF	Point coordination function
PD-NOMA	Power-domain NOMA
PIC	Parallel interference cancellation
PLC	Programmable logic controllers
PLS	Physical layer security
PPDU	Physical layer protocol data unit
PPG	Photo plethysmo gram
PPM	Pulse position modulation
PPM	Pulse position modulation
PQC	Post-quantum cryptography
PSO	Particle swarm optimization
PTP	Precision time protocol
QAM	Quadrature amplitude modulation
QC	Quantum computing
QoS	Quality of Service
QSC	Quantum-safe cryptography
RAN	Radio access network
RAW	Reliable and available wireless
RIFS	Reduced interframe spacing
RIS	Reconfigurable intelligent surfaces
RL	Reinforcement learning
RNN	Recurrent neural network

RoHS	Regulation of Hazardous Substances
RS	Reconciliation sub-layer
RU	Resource unit
SC	Superposition coding
SDF	Specification document framework
SDN	Software-defined network
SIC	Successive interference cancellation
SIDH	Singular isogeny Diffie–Hellman
SIFS	Short inter-frame space
SN	Serving network
SNMP	Simple network management protocol
SOP	Secrecy outage probability
SPEC	Standard Performance Evaluation Corporation
SU	Secondary users
SUCI	Subscription concealed identifier
SUMO	Simulation of urban mobility
SVM	Support vector machine
TAS	Time-aware scheduling
TC	Trafıñ$\mathcal{A}$c class
TDD	Time division duplex
TF	Transformation matrix
TF	Trigger frame
TPR	Truncated pyramid arrangement
TS	Trafıñ$\mathcal{A}$c stream
TSN	Time sensitive networking
TSPEC	Trafıñ$\mathcal{A}$c speciïñ$\mathcal{A}$cation
TWT	Target wake time
TxOP	Transmission opportunity
UCGD	Uniform channel gain difference
UCW	Uora contention window
UE	User entity
UI	User information
UL	Uplink
UNECA	United Nations Economic Commission for Africa
URLLC	Ultra-reliable low–latency communication
V2X	Vehicle-to-everything
VM	Virtual machine
vROA	Variable receiver orientation angle

VROA	Variable receiver orientation angle
WEEE	Waste electrical and electronic equipment
WMM	Wi-Fi multi media (Wi-Fi Alliance Version of EDCA)
WMM-SA	Wi-Fi multi media scheduled access (Wi-Fi Alliance Version of HCCA)
WP	Working party
XGBoost	Extreme gradient boosting
ZF	Zero forcing

1

Intelligent Security for DDoS in HetIoT (6G Perspective)

Shalaka S. Mahadik[1], Pranav M. Pawar[2], Raja M.[3], Dnyaneshwar Mantri[4], Neeli Rashmi Prasad[5], and Nandkumar P. Kulkarni[6]

[1,2,3]Birla Institute of Technology Pilani, UAE
[4]Sinhgad Institute of Technology, India
[5]TrustedMobi "VehicleAvatar Inc.," USA
[6]NIT-ADT University, India

Abstract

The heterogeneous Internet of Things (HetIoT) infrastructure is a prime target for attackers, due to a lack of cybersecurity measures. The HetIoT devices contain various security holes because of its heterogeneous nature, and attack tactics are advancing dramatically as the world transforms from 5G to 6G technology. The attackers can easily infiltrate the HetIoT infrastructure and form botnets to launch distributed denial of service (DDoS) attacks. Heading towards 6G technology, its speed, and connectivity provide a faster communication platform for various IoT devices; however, the number of vulnerabilities and overall records vulnerable to exploits will also increase. The explosive growth of 6G connectivity and HetIoT will make spotting and blocking of DDoS attacks challenging in the upcoming era. Hence, it is necessary to spot and block the attacks intelligently using different artificial intelligence-based learning techniques. The design of intelligent security (IS) techniques for addressing security and privacy issues in HetIoT is gaining popularity, due to its efficiency in detection and prediction of attacks. Incorporating security with the 6G technology in HetIoT using the IS approach will result in a promising solution. Therefore, the chapter concentrates on protecting HetIoT infrastructure against DDoS threats using IS approaches. Additionally, the article sheds light on the current developments in IS approaches and their open challenges, issues, and solutions regarding the 6G perspective.

Keywords: Intelligent Security, DDoS, 6G, Heterogeneous Internet of Things

1.1 Introduction

The IoT is an advancing technology that is gaining popularity yearly in diverse domains such as education, robotics, agriculture, and Industry 4.0, to name a few. This new revolution requires upgrading communication networks, speed, connectivity, reliability, and coverage as the HetIoT-connected devices and sensors are snowballing, and by 2025, it will go beyond 17 million [1]. Although the current 5G technology is mature enough, the continuous development of HetIoT application domains such as autonomous driving and baby monitoring cameras demands higher data transfer rates and ultra-low latency, which 5G fails to fulfill [2]. According to a scientific report, the 6G technique enhances the performance and latency by utilizing the distributed radio access network and terahertz frequency [3, 4, 5]. The operational 6G technology is envisaged in 2030. As a result, 6G technology can be the facilitator for such high-demand HetIoT application domains. With 6G-enabled HetIoT infrastructure, billions of IoT devices and applications can interact at extremely high data rates to meet the demands of the fully connected digital community [6]. On the one hand, integrating 6G into HetIoT infrastructure can be beneficial. Still, it also brings new challenges, such as network mobility, energy consumption, resource management, reliable and consistent connectivity, and security and privacy [5, 7, 8, 9]. Figure 1.1 exhibits the significant challenges arising from the 6G-enabled HetIoT infrastructure.

The 6G connectivity allows the HetIoT infrastructure to disperse worldwide faster with remote access, exposing it to various security and privacy vulnerabilities. The HetIoT devices hold vital information about their users; for example, smart watches store information about a patient's geolocation and health records in the medical field. As a result, protecting the HetIoT environment from cyber threats such as scalable remote attacks, cryptographic side-channel attacks, DDoS attacks, data breaches, malware, and other threats is a major concern while acquiring the 6G technology. Moreover,

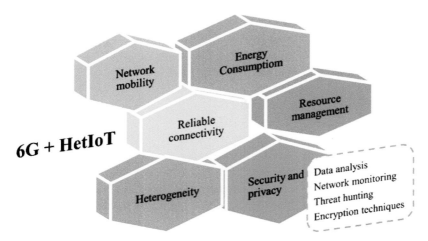

Figure 1.1   Challenges in 6G-enabled HetIoT infrastructure.

the perpetrators use more sophisticated approaches, embracing a significant risk in the HetIoT ecosystem [10].

The DDoS attack is one of the most influential and popular security threats in the HetIoT infrastructure that entails hijacking the HetIoT devices to drain resources and disrupt services [1, 11]. This attack will explicitly tear down the application hosting the services to which the users are trying to connect, or it may flood the network bandwidth, resulting in network failure and a server outage. The future 6G network is more open and heterogeneous than 5G [12]. Additionally, as the 6G is an open and integrated space-air-ground network, traditional border security solutions such as firewalls and intrusion detection systems (IDSs) may provide an inadequate defense to the HetIoT architecture against the DDoS threat. More elastic security techniques must be devised to meet the 6G objectives in the HetIot environment. According to recent AT& T Cybersecurity Intelligence research, DDoS attacks will "hit new peaks" as the availability of HetIoT devices using 6G connection increases. It gives more chances for nefarious actors to exploit and consume more bandwidth at their disposal [13].

Applying intelligence to security has proven its effectiveness in safeguarding the HetIoT ecosystems [4, 11, 12, 14]. The main reason behind the success of intelligent security (IS) techniques, including machine learning (ML) and deep learning (DL), is the abundance of data, i.e., the HetIoT system generates a huge amount of data, which is feasible to develop the various IS algorithms from a security standpoint. While developing the IS technique, it fed the data to analyze the anomalous network traffic during training and provided the best results for detecting security threats. The combination of HetIoT, 6G, and IS will be the most beneficial way to assess the complicated security and privacy issues while also helping to reduce the redundant data used in mass communication.

With the above-mentioned related study, this chapter is motivated to focus on the security of the HetIoT from a 6G perspective by mainly focusing on DDoS attacks. The key contribution of this chapter is to highlight the impact of the DDoS threat on the HetIoT infrastructure from a 6G viewpoint. The chapter details the various types of DDoS attacks present and the current state-of-the-art work done so far to safeguard the HetIoT using IS techniques, namely, ML, DL, reinforcement learning (RL), and federated learning (FL). The chapter also includes the open challenges, issues, and solutions in safeguarding the HetIoT infrastructure from a 6G viewpoint.

The remainder of this chapter is organized as follows: Section 1.2 overviews DDoS attacks and their types. Section 1.3 focuses on the current state-of-the-art work using various IS techniques from a HetIoT perspective. Section 1.4 includes the open challenges and issues in IS techniques. Finally, the last section ends with a conclusion.

1.2 DDoS in HetIoT with 6G Perspective

As the technology is upgraded, the trend of DDoS attacks is also growing, and the same thing has been highlighted by Cisco in its annual report [15]. A DDoS attack is launched by a network of infected devices known as the botnet. The perpetrators take control of such devices and utilize them to attack legitimate users' devices. As per the related study, the various types of DDoS attacks are highlighted in Figure 1.2, including flooding-based, protocol-based, and reflection- or amplification-based attacks.

Figure 1.3 displays a DDoS threat example in HetIoT infrastructure. Due to the traditional threat vectors like weak passwords, open ports, etc., the perpetrator succeeds in taking control of various IoT devices, forming a DDoS botnet. The perpetrator targets the DNS server, floods the network with UDP traffic, and makes the server unavailable to legitimate users.

The HetIoT devices are resource-constrained, i.e., low-power, low-cost, and have limited computational capacity, making them more brittle and challenging in the 6G-enabled HetIoT context. Most conventional solutions for mitigating DDoS attacks on communications networks are nearly obsolete and must be revised to meet today's escalating security

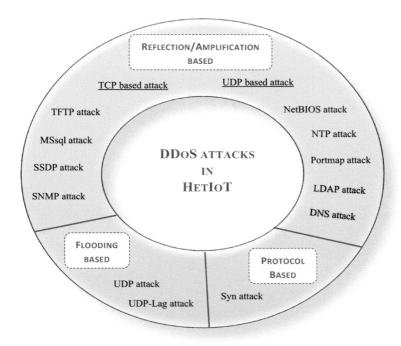

Figure 1.2   Types of DDoS threat in HetIoT [11].

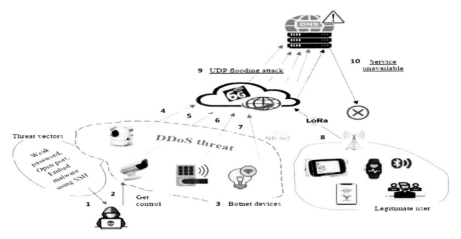

Figure 1.3   DDoS threat on HetIoT infrastructure.

requirements. However, the various models developed using IS techniques exhibit their aptitude to protect the HetIoT applications with higher accuracy. The resource-constrained aspect of the HetIoT applications demands the designing of light-weight cryptographic techniques, lightweight IS models regarding the number of layers and neurons, trust-based network architecture, and lower computation overhead, which is a challenging task and needs to be handled carefully while developing better IS model. Additionally, as mentioned in [16], another challenge in developing IS techniques is the availability of the datasets. The heterogeneous network architecture needs to be considered while creating the dataset for DDoS attack detection.

1.3 State-of-the-art Work: Intelligent Security (IS)

With IS algorithms, recognizing DDoS activity early and conducting swift, targeted, and optimized abatement operations to stop such attacks is possible nowadays [17]. The section investigates recent state-of-the-art work using IS techniques from 2020 to 2023 to identify and mitigate the DDoS attack concerning ML, DL, RL, and FL, illustrated in Figure 1.4.

1.3.1 ML-based technique

The author in [18] addresses the thresholding and ML approach issue by utilizing an extreme gradient boosting (XGBoost) algorithm and adaptive bandwidth control approach that help reduce the packet drop ratio and improve the threat detection accuracy. The author in [19] investigates the overall vulnerabilities of IoT systems to cyber threats using ML approaches, with a focus on DoS attacks. The paper assesses the CICDDoS2019 dataset

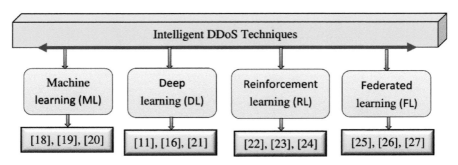

Figure 1.4   State-of-the-art IS techniques to protect DDoS threat.

using the logistic regression (LR) algorithm and achieves a reasonable accuracy rate. The paper focuses on benign, LDAP, and NetBIOS attacks, whereas many other attacks in the dataset are not considered. Software-defined network (SDN) offers less computation overhead on IoT network devices; hence, the author in [20] proposed an adaptive ML framework for detecting DDoS attacks by simulating its dataset.

1.3.2 DL-based technique

The heterogeneous network of IoT makes designing security models difficult; however, DL approaches can handle such networks and detect DDoS attacks with high accuracy. As a result, the author in [16] suggests a deep belief network (DBN) technique for DDoS threat screening. The author in [21] focuses on cloud service protection using long short-term memory (LSTM) against DDoS threats. The author in [11] proposed an IDS using a convolutional neural network (CNN) approach for DDoS attack detection and classification in heterogeneous IoT environments using the CICDDoS2019 dataset that holds all recent DDoS attacks

1.3.3 RL-based technique

The author in [22] addresses the issue of the existing IS techniques that mainly focus on high-rate DDoS threat detection and proposed slow-rate DDoS threat detection and mitigation techniques using deep reinforcement learning (DRL). The performance of the proposed approach is evaluated using simulated flow-based traffic in an SDN context. The author in [23] proposed a moving target defense (MOTAG) system using the deep Q-learning (DQ) technique and highlighted that the proposed methods actively shuffle the defense system with less network resource consumption. With the focus on Internet-connected fast-moving vehicles to detect DDoS threats, author [24] proposes the RL based algorithm using q-learning and double q-learning network (DDQN). The proposed approach monitors the changes in data traffic when the vehicles are moving from one base station to another; this drastic change in the traffic information is used to detect benign and DDoS threats.

1.3.4 FL-based technique

FL techniques help to maintain users' data privacy by utilizing a distributed training approach. The author in [25] proposed an FL approach using CNN to detect DDoS threats. The author in [26] utilized autoencoder (AE) for feature extraction and recurrent neural network (RNN) for the detection of DDoS attacks. The proposed approach employed K-means and SMOTE techniques to handle the data imbalance issue. While preserving data privacy using the blockchain technique, the author in [27] proposed a trusted-based multi-domain protection system to detect DDoS threats.

The comparative analysis of the state-of-the-art work using ML, DL, RL, and FL techniques reviewed concerning the IS algorithm, threats address, dataset utilized, evaluation metrics, and experimental setup employed are highlighted in Table 1.1. Table 1.1 provides a roadmap for new academics or researchers, highlighting the most recent state-of-the-art work using various IS techniques. It also helps to understand that the majority model focuses only on binary classification, i.e., benign and DDoS threats. Different DDoS threats are present in reality and must be dealt with as the 6G connectivity makes the HetIoT ecosystem more threatening. Further, Table 1.1 outlines the availability of the dataset, performance metrics, and experimental setup necessitated while designing any IS algorithm.

1.4 Open Issues, Challenges, and Solution in IS (DDoS)

HetIoT applications employ various types of devices, various modes of communication, and a wide range of data being communicated. The literature review shows that IS techniques improve performance and accuracy when handling massive data while reducing false alarm rates for detecting DDoS threats. In light of the large amount of data produced, transmission overhead, and data privacy concerns are addressed by spreading traffic data across multiple network devices that must be clustered together, resulting in the challenges experienced during IS model training [25, 28]. The article highlighted a few challenging issues while adopting IS approaches in conjunction with 6G and HetIoT contexts.

1.4.1 Heterogeneous network architecture

Heterogeneous networks have gained considerable attention in recent decades as they possess the opportunity to be an innovative framework for Darwinian networks and one of the design issues in 6G technology [6]. Applications like virtual reality and smart electric vehicles are instances of such network architecture. Regarding the HetIoT infrastructure, there is a lack of a standard design for such networks and devices. Different vendors have

Table 1.1 Comparative analysis of the state-of-the-art work

Year	Algorithm	Threats	Dataset	Evaluation metrics	Experimental set-up	Application domain
				ML techniques		
2020	Xgboost [18]	Benign, DDoS attack	CICDDoS2019, NSL-KDD, CAIDA	Acc, Pr, Recall, F1-score, FPR	Mininet, Hping3, iperf3 tool	SDN
2021	LR [19]	Benign, LDAP, NetBIOS	CICDDoS2019	Accuracy (Acc)	Python 3.7, Scikit-learn	IoT
2022	SVM, NB, KNN, LR,RF [20]	Benign, DDoS attack	Simulated	Acc, Pr, Recall, F1-score	Oracle VirtualBox, Mininet, tshark, Python	SDN
				DL techniques		
2021	DBN [16]	Benign, DDoS attack	Simulated	Acc, Pr, Recall, F1-score	TensorFlow python and Scikit-learn	Heterogeneous IoT
2022	LSTM [21]	Benign, MSSQL, SYN, UDP	CICDDoS2019	Acc, training and testing time, RMSE	Google collab with GPU and TPU	Cloud services
2023	CNN [11]	Multiclass	CICDDoS2019	Acc, Pr, Recall, F1-score	TensorFlow python, Scikit-learn, Matplotlib	Heterogeneous IoT

(Continued)

Table 1.1 *Continued.*

RL techniques							
2020	DQ-MOTAG [23]	Benign, attack	DDoS	Simulated	Error block rate, Network resource consumption	jikecloud servers, TensorFlow 1.14.0, Python 3.6	IoT
2021	Q-learning and DDQN [24]	Benign, attack	DDoS	Real-time	Accuracy, F1-Score	OpenStreetMap, ddosflowgen and hping3 tool	IoVehicles
2022	DRL [22]	Benign, attack	DDoS	Simulated	Average detection rate, Acc	Mininet, ONOS controller, Apache web server	SDN
FL techniques							
2021	FL-CNN [25]	Binary and Multiclass		CICDDoS2019	Acc, Pr, Recall, F1-score	PyTorch, Python 3.8, Anaconda IDE	IoT
2021	AE, RNN [26]	Multiclass		CICDDoS, CICIDS, NLSKDD	Acc, Pr, Recall, F1-score	PyTorch, Python 3.8	IoT
2022	HomoCNN [27]	Multiclass		Simulated	Acc, Pr, Recall, F1-score	Python scapy library, Vmware, FATE framework	Blockchain and IoT

different architectures and protocols; hence, the security suffers the most in this heterogeneous network. The data generated from such a network will be noisy, inaccurate, and incomplete. Such data may result in poor accuracy and a high false alarm rate while training any IS algorithm. Therefore, change in network architecture causes challenging issues while designing any IS technique, which needs attention.

1.4.2 Intelligent edge computing

Another current area of research that seeks remedies to the limiting nature of HetIoT technology is edge computing. With the integration of 6G connectivity and HetIoT, data must travel to the cloud worldwide, causing privacy and trust issues. Edge computing employs IS techniques to conduct computation operations at the network's edge, resulting in lower latency and the preservation of user data privacy. Compared to cloud services, edge devices have less

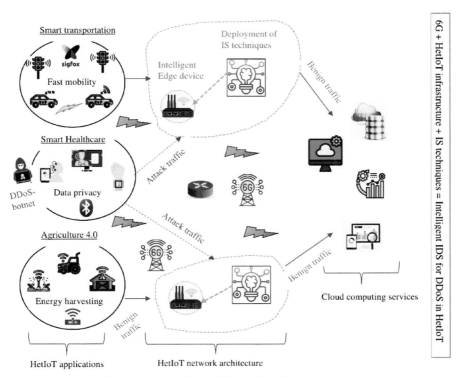

Figure 1.5   Intelligent IDS for DDoS threat defense.

computation power; hence, they demand the development of lightweight IS models. The author in [1] proposes the smart IDS using an LSTM-based DL technique, which is lightweight and less complex than identifying the DDoS attack in HetIot infrastructure. Such intelligent IDS can be easily deployed at the network's edge to successfully detect the various types of DDoS attacks. As a result, developing a lightweight, intelligent IDS can solve IS challenges in the future 6G-enabled HetIoT context while maintaining privacy and trust. This innovation will enhance security in healthcare and baby monitoring applications. The deployment of intelligent IDS in the future 6G and HetIoT environments is depicted in Figure 1.5.

1.4.3 Smart federated learning

Although in the current era, federated learning is an excellent solution to mitigate the security threat in the HetIoT context, it faces two challenges while preserving the privacy of data, i.e., model poisoning and reverse engineering, which is an active research area. Quantum cryptography can be the solution to safeguard the user's privacy. This technique needs to be developed more; its key distribution, long-distance traveling, and cost are challenging issues and need researchers' attention.

1.5 Conclusion

The chapter exhibits the demand for transitioning from 5G to 6G connectivity, which would help the HetIoT infrastructure's rapid growth. The 6G connectivity allows the HetIoT infrastructure to disperse over the world much faster, exposing its components to significant security and privacy vulnerabilities. The chapter solely focuses on the impact of DDoS threat in the HetIoT environment with the 6G viewpoint. An efficient and intelligent intrusion detection is an anomaly detection system that thoroughly monitors network behavior and provides adequate results with high accuracy and low false alarms. The chapter highlights the IS approaches projected to assist 6G-enabled HetIoT infrastructure in making more optimal and adaptive decisions. The chapter also outlines the open, challenging issues and solutions to protect the future 6G and HetIot infrastructure against the DDoS threat.

References

[1] Mahadik Shalaka S., Pranav M. Pawar, Raja Muthalagu, Neeli R. Prasad, and Dnyaneshwar Mantri, "Smart LSTM-based IDS for Heterogeneous IoT (HetIoT)", In 2022 25th International Symposium on Wireless Personal Multimedia Communications (WPMC), pp. 23-28, IEEE, 2022.

[2] Asma Alotaibi, Ahmed Barnawi, "Securing massive IoT in 6G: Recent solutions, architectures, future directions", Elsevier, Internet of Things, Volume 22, 2023, 100715, ISSN 2542-6605.https://doi.org/10.1016/j.iot.2023.100715.

[3] Mahdi MN, Ahmad AR, Qassim QS, Natiq H, Subhi MA, Mahmoud M., "From 5G to 6G Technology: Meets Energy, Internet-of-Things and Machine Learning: A Survey", Applied Sciences, 2021, 11(17):8117.https://doi.org/10.3390/app1117811.

[4] Zakria Qadir, Khoa N. Le, Nasir Saeed, Hafiz Suliman Munawar, "Towards 6G Internet of Things: Recent advances, use cases, and open challenges", ICT Express, 2022, ISSN 2405-9595.https://doi.org/10.1016/j.icte.2022.06.006.

[5] Y. Siriwardhana, P. Porambage, M. Liyanage and M. Ylianttila, "AI and 6G Security: Opportunities and Challenges," 2021 Joint European Conference on Networks and Communications & 6G Summit (EuCNC/6G Summit), Porto, Portugal, 2021, pp. 616-621.https://doi:10.1109/EuCNC/6GSummit51104.2021.9482503.

[6] Kulkarni, Nandkumar P., Dnyaneshwar S. Mantri, Neeli R. Prasad, Pranav M. Pawar, and Ramjee Prasad, "6G Future Vision: Requirements, Design Issues and Applications", In 6G Enabling Technologies, pp. 23-43, River Publishers, 2023.

[7] K. K. Vaigandla, "Communication Technologies and Challenges on 6G Networks for the Internet: Internet of Things (IoT) Based Analysis," 2022 2nd International Conference on Innovative Practices in Technology and Management (ICIPTM), Gautam Buddha Nagar, India, 2022, pp. 27-31.https://doi:10.1109/ICIPTM54933.2022.9753990.

[8] Alhashimi, Hayder Faeq, MHD Nour Hindia, Kaharudin Dimyati, Effariza Binti Hanafi, Nurhizam Safie, Faizan Qamar, Khairul Azrin, and Quang Ngoc Nguyen, "A Survey on Resource Management for 6G Heterogeneous Networks: Current Research, Future Trends, and Challenges", Electronics, 2023, 12, no. 3: 647.https://doi.org/10.3390/electronics12030647.

[9] Abdel Hakeem SA, Hussein HH, Kim H., "Security Requirements and Challenges of 6G Technologies and Applications", Sensors (Basel), 2022 March 2;22(5):1969.https://doi:10.3390/s22051969.PMID: 35271113;PMCID:PMC8914636.

[10] Roland Atoui, "Securing IoT with Quantum Cryptography," URL=http s://www.iotforall.com/securing-iot-with-quantum-cryptography, [Online; accessed January 14, 2022].

[11] Mahadik Shalaka S., Pawar P.M. and Muthalagu R., "Efficient Intelligent Intrusion Detection System for Heterogeneous Internet of Things (HetIoT)," J Netw Syst Manage, Springer, 31, 2, 2023.https://doi: 10.1007/s10922-022-09697-x.

[12] Chen, Xu, Wei Feng, Ning Ge, and Yan Zhang, "Zero trust architecture for 6G security", 2022, arXiv preprint arXiv:2203.07716.

[13] Catherine Sbeglia Nin, "DDoS trend volumetric attacks", URL=https: //www.rcrwireless.com/20230214/security/ddos-trends-volumetric-att acks-are-on-the-rise,[Online;accessed-February14,2023].

[14] Joseph, Diana Susan, Pranav M. Pawar, and Rahul Pramanik, "Intelligent plant disease diagnosis using convolutional neural network: a review," Multimedia Tools and Applications, pp. 1-67, 2022.

[15] Cicsco DDoS Annual Report (white paper), URL=https://www.cisco.co mlclenlus/solutions/collaterallexecutiveperspectiveslannual-inte,[Onli ne;accessedMarch9,2020].

[16] Amaizu GC, Nwakanma CI, Bhardwaj S, et al., "Composite and Efficient DDoS attack Detection framework for B5G networks", Computer Networks, Elsevier, vol. 188:107, pp.871, 2021.

[17] Alex Pavlovic, "How AI/ML Can Thwart DDoS Attacks," URL=https: //www.darkreading.com/dr-tech/how-ai-ml-can-thwart-ddos-attacks, [Online; accessed December 20, 2022].

[18] H. A. Alamri, V. Thayananthan, "Bandwidth control mechanism and extreme gradient boosting algorithm for protecting software-defined networks against DDoS attacks," IEEE Access 8, IEEE, 2020, pp.194269-194288.

[19] S. Chesney, K. Roy, S. Khorsandroo, "Machine learning algorithms for preventing IoT cybersecurity attacks," Proceedings of SAI Intelligent Systems Conference, Springer, 2020, pp. 679-686.

[20] Aslam M, Ye D, Tariq A, Asad M, Hanif M, Ndzi D, Chelloug SA, Elaziz MA, Al-Qaness MAA, Jilani SF., "Adaptive Machine Learning

Based Distributed Denial-of-Services Attacks Detection and Mitigation System for SDN-Enabled IoT", Sensors, 2022; 22(7):2697.https://doi.org/10.3390/s22072697.

[21] Aydın, Hakan, Zeynep Orman, and Muhammed Ali Aydın. "A long short-term memory (LSTM)-based distributed denial of service (DDoS) detection and defense system design in public cloud network environment." Computers & Security 118, 2022: 102725.

[22] Yungaicela-Naula, Noe M., Cesar Vargas-Rosales, Jesús Arturo Pérez-Díaz, and Diego Fernando Carrera, "A flexible SDN-based framework for slow-rate DDoS attack mitigation by using deep reinforcement learning," Journal of Network and Computer Applications, 205, 2022: 103444.

[23] Li, Zhong, Yubo Kong, Cheng Wang, and Changjun Jiang, "DDoS mitigation based on space-time flow regularities in IoV: A feature adaption reinforcement learning approach," IEEE Transactions on Intelligent Transportation Systems 23, no. 3, 2021, pp. 2262-2278.

[24] Chai, Xinzhong, Yasen Wang, Chuanxu Yan, Yuan Zhao, Wenlong Chen, and Xiaolei Wang, "DQ-MOTAG: deep reinforcement learning-based moving target defense against DDoS attacks," In 2020 IEEE Fifth International Conference on Data Science in Cyberspace (DSC), pp. 375-379. IEEE, 2020.

[25] D. Lv, X. Cheng, J. Zhang, W. Zhang, W. Zhao, and H. Xu, "DDoS Attack Detection Based on CNN and Federated Learning," 2021 Ninth International Conference on Advanced Cloud and Big Data (CBD), Xi'an, China, 2022, pp. 236-241.https://doi:10.1109/CBD54617.2021.00048.

[26] J. Zhang, P. Yu, L. Qi, S. Liu, H. Zhang, and J. Zhang, "FLDDoS: DDoS Attack Detection Model based on Federated Learning," 2021 IEEE 20th International Conference on Trust, Security, and Privacy in Computing and Communications (TrustCom), Shenyang, China, 2021, pp. 635-642.https://doi:10.1109/TrustCom53373.2021.00095.

[27] Yin, Ziwei, Kun Li, and Hongjun Bi., "Trusted Multi-Domain DDoS Detection Based on Federated Learning," Sensors 22, no. 20, 2022: 7753.

[28] Mahadik, Shalaka S., Pranav M. Pawar, and Raja Muthalagu., "Edge-HetIoT Defense against DDoS attack using Learning Techniques," Computers & Security (2023): 103347.

Biographies

Shalaka Shankar Mahadik is a Ph.D. Researcher at BITS Pilani, Dubai campus, UAE. Her research interest includes computer networks, information security, Internet of Things (IoT), and mobile communication. She completed her B.E. and M.E. (Computer Engineering) from Mumbai University (India) in 2005 and 2011, respectively. She worked as a lecturer from 2005 to 2011 and as an assistant professor from 2011 to 2014. She has participated in and organized various workshops and national conferences. Her research is currently based on security and privacy in heterogeneous IoT (HetIoT). She published her research work in well-known, reputed conferences and journals like IEEE, Springer, and Elsevier.

Pranav M. Pawar graduated in Computer Engineering from Dr. Babasaheb Ambedkar Technological University, Maharashtra, India, in 2005, received a Master in Computer Engineering from Pune University, in 2007, and received Ph.D. in Wireless Communication from Aalborg University, Denmark in 2016; his Ph.D. thesis received a nomination for Best Thesis Award from Aalborg University, Denmark. Currently, he is working as an assistant professor in the Dept of Computer Science, Birla Institute of Technology and Science, Dubai. Prior to BITS he was a postdoctoral fellow at Bar-Ilan University, Israel from March 2019 to October 2020 in the area of Wireless Communication and Deep Leaning. He is the recipient of an outstanding postdoctoral fellowship from the Israel Planning and Budgeting Committee. Dr. Pawar worked as an Associate Professor at MIT ADT University, Pune from 2018 to 2019 and also as an associate professor in the Department of Information Technology, STES's Smt. Kashibai Navale College of Engineering, Pune from 2008 to 2018. From 2006 to 2007, he was working as System Executive in POS-IPC, Pune, India. He received recognition from Infosys Technologies Ltd. for his contribution to the Campus Connect Program and also received different funding for research and attending conferences at the international level. He published more than 40 papers at the national and international levels. He is IBM DB2 and IBM RAD certified professional and completed

NPTEL certification in different subjects. His research interests are energy-efficient MAC for WSN, QoS in WSN, wireless security, green technology, computer architecture, database management system, and bioinformatics.

Raja Muthalagu is currently an Associate Professor with Birla Institute of Technology and Science, Pilani, Dubai Campus, Dubai, USA. He was a Postdoctoral Research Fellow with Air Traffic Management Research Institute, Nanyang Technological University, Singapore, from 2014 to 2015. Dr. Muthalagu was the recipient of the Canadian Commonwealth Scholarship Award 2010 for the Graduate Student Exchange Program in the Department of Electrical and Computer Engineering, University of Saskatchewan, Saskatoon, SK, Canada. His research interest includes wireless communication, signal processing, aeronautical communication, and cybersecurity.

Dnyaneshwar S Mantri graduated in Electronics Engineering from Walchand Institute of Technology, Solapur (MS) India in 1992 and received Master's from Shivaji University in 2006. He has been awarded Ph.D. in Wireless Communication at the Center for TeleInFrastruktur (CTIF), Aalborg University, Denmark. He has teaching experience of 25 years. From 1993 to 2006 he was working as a lecturer in different institutes [MCE Nilanga, MGM Nanded, and STB College of Engg. Tuljapur (MS) India]. Since 2006 he is associated with Sinhgad Institute of Technology, Lonavala, Pune and presently he is working as a professor in the Department of Electronics and Telecommunication Engineering. He is a recognized master's and Ph.D guide at Savitribai Phule Pune University, Pune in the subjects of Electronics and Telecommunication Engg, Computer and Information Technology. He is a Senior Member of IEEE, a Fellow of IETE, and Life Member of ISTE. He has published 06 books, 20 journal papers in indexed and reputed journals (Springer, Elsevier, IEEE, etc.), and 19 papers in IEEE conferences. He is a reviewer of international journals (Wireless Personal Communication, Springer, Elsevier, IEEE, Communication Society, MDPI, etc.) and conferences organized by IEEE. He worked as TPC member for various IEEE conferences and also organized IEEE conferences GCWCN2014 and GCWCN2018. He worked on various committees at University and College. He was a member of the Board of Studies in Electronics at Dr. Babasaheb Ambedkar Marathwada University, (Dr.BAMU) Aurangabad, He is a guest editor in STM journals, elected as Executive Council member (EC), and vice chair of IETE Pune Local Center (2022-24). His research interests are in Adhoc Networks, Wireless Sensor Networks, Wireless Communications specific focus on energy and bandwidth.

 **Ir. Neeli R. Prasad**, CTO of SmartAvatar B.V. Netherlands and VehicleAvatar Inc. USA, IEEE VTS Board of Governor Elected Member & VP Membership. She is also full professor at Department of Business Development and Technology (BTech), Aarhus University. Dr. Neeli is a cybersecurity, networking, and IoT strategist. She has throughout her career been driving business and technology innovation, from incubation to prototyping to validation and is currently an entrepreneur and consultant in Silicon Valley. She has made her way up the "waves of secure communication technology" by contributing to the most ground breaking and commercial inventions. She has general management, leadership and technology skills, having worked for service providers and technology companies in various key leadership roles. She is the advisory board member for the European Commission H2020 projects. She is also a vice chair and patronage chair of IEEE Communication Society Globecom/ICC Management & Strategy Committee (COMSOC GIMS) and Chair of the Marketing, Strategy and IEEE Staff Liaison Group. Dr. Neeli Prasad has led global teams of researchers across multiple technical areas and projects in Japan, India, throughout Europe and USA. She has been involved in numerous research and development projects. She also led multiple EU projects such as CRUISE, LIFE 2.0, ASPIRE, etc. as project coordinator and PI. She has played key roles from concept to implementation to standardization. Her strong commitment to operational excellence, innovative approach to business and technological problems and aptitude for partnering cross-functionally across the industry have reshaped and elevated her role as project coordinator making her a preferred partner in multinational and European Commission project consortiums. She has four books on IoT and Wi-Fi, many book chapters, peer-reviewed international journal papers and over 200 international conference papers. Dr. Prasad received her Master's degree in electrical and electronics engineering from Netherland's renowned Delft University of Technology, with a focus on

personal mobile and radar communications. She was awarded her Ph.D. degree from Università di Roma "Tor Vergata", Italy, on Adaptive Security for Wireless Heterogeneous Networks.

 **N. P. Kulkarni** received Bachelor of Engineering (B.E.) degree from Walchand College of Engineering, Sangli, Maharashtra, (India) in 1996. He has been with Electronica, Pune from 1996 -2000. He worked on retrofits, CNC machines and was also responsible for PLC programming. In 2000, he received the Diploma in Advanced Computing (C-DAC) degree from MET's IIT, Mumbai. In 2002, He became Microsoft Certified Solution Developer (MCSD). He worked as a software developer and system analyst in CITIL, Pune and INTREX India, Mumbai respectively. He has 23 years of experience both in industry and academia. From 2002 onwards he is working as a faculty in Savitribai Phule Pune University, Pune. Since 2007, he was working with SKNCOE, Pune as a faculty in IT Department. Presently working as an associate professor at the School of Computing, MIT ADT University, Pune. He completed a Master of Technology (M. Tech) degree with computer specialization from COEP, Pune (India) in 2007 and Ph.D. from Aarhus University, Denmark in 2019. His area of research is in WSN, VANET, and Cloud Computing. He has published papers in 18 international journals, 15 International IEEE conferences, and three national conferences.

2

Industry 5.0 and 6G: Human-centric Approach

Dnyaneshwar S. Mantri[1], Pranav M. Pawar[2], Nandkumar P. Kulkani[3], Neeli R. Prasad[4], and Ramjee Prasad[5]

[1]Sinhgad Institute of Technology, India
[2]BITS Pillani, Dubai
[3]MIT ADT University, India
[4]TrustedMobi "Vehicle Avatar Inc.," USA
[5]Department of Business Development and Technology, Aarhus University, Denmark

Abstract

Technological improvements and evolutions are required beyond fifth-generation (5G) networks for wireless communications as well as in the industry. The involvement of collaborative robots (COBOT) will satisfy the personal needs of human beings as and when required leading to Industry 5.0. Industrial revolutions relate to the human–machine interactions to satisfy needs in an easy and quicker way. It also helps to convert your ideas into implementations with the help of future-generation networks such as 6G. The trillions of connected devices would demand exceptionally high-performance, interconnectivity, especially in dynamic circumstances such as varied mobility, extremely high density, and vibrant surroundings. The mass personalization in Industry 5.0 is supported by the new advent of technology support as artificial intelligence (AI) technique, big data, cloud computing (CC), and robotics, i.e., smart cyber physical cognitive systems (CPCS). The enabling technologies such as 6G may be a scalable, adaptable, and long-lasting wireless access mechanism to handle a wide range of requirements, value chains, supply chains, research, and innovations in the new design of business models for attaining the highest efficiency. This chapter examines the essential requirements, and enabling technologies used in Industry 5.0 with 6G evolutions. In order to accomplish the aim, it put forwards a proposal for enhancement of the 4.0 to 5.0 vision of industrial revolutions using next-generation networks and technologies such as CC, Robotics, AI, IoT, etc.

Keywords: Industrial Revolution Mass Personalization, Artificial Intelligence, Internet of Things (IoT), 6G, Network, Industry 5.0, Smart Factory

2.1 Introduction

Over the last decade, an unprecedented convergence in the various technology domains raised expectations not only in the automation industry but also in social means. The cultivated extreme automation, hyperconnectivity, technology enhancement, birth of various sensors, big data analytics, information communication technology (ICT), collaborative robots (COBOTS) powered by AI and machine learning, and Internet of Things permit real-time data analysis and control actions in the Industry 5.0. Looking back into the industrial revolutions from Industry 1.0 to Industry 4.0, the drastic changes that happened is in technology moving from mechanization to virtualization and interconnections with Internet to cyber−physical systems. The objective of these revolutions is to create a connected world in which everyone and everything is linked, i.e., personalization. Smooth communication and interaction between various objects, people, machine, and device is feasible due to the availability of Internet. Also, 5G networks can offer mobile broadband connectivity, reliable and massive MIMO-type communication with low latency in order to improve QoS together with network performance. The latest 6G developments and deployments can provide high-quality service achieving ultra-low latency and the highest data rates (Tbps).

Industry 5.0 supports for development of new designs in business models, value and supply chains, in digital transformation, formulating new public policy for collaboration, tools for innovation in research, as well as vertical and horizontal coherence through the use of international standards and action at all levels of government. In all the application cases mentioned, if networks utilized for Industry 4.0 are tried to be implemented for Industry 5.0, then they will be complicated and chaotic. The Industry 5.0 Revolution must take place in market with sophistication and easy to provide clients a variety of new services and vertical applications. For that, AI, big data, and computing will become obligatory and indispensable elements of the 6G network. In today's world, virtually every industry recognizes the importance of big data and is being used to boost many industries and technological advancements. Recently, collaborative robots working on the human–machine re-emergence, founded upon big data, AI/ML algorithms, and human bond communications have generated new possibilities of research in cyber–physical cognitive systems.

The contribution of this chapter is to raise awareness of the most recent technological advancements and obstacles in enabling Industry 5.0 networks. The chapter also discusses the integration and convergence of new technologies (ICT, IoT, cloud computing, big data, and robotics), allowing for the provision of 6G services at anytime, anyplace, and anything to become personal. Industry 5.0 helps to interact with humans and machines directly to get the best services at lower rates and in quicker time.

In this chapter, we give a thorough review of Industrial Revolution, analyze the use of leading-edge technologies to transform Industry 4.0 to 5.0 Revolutions, and talk about a number of outstanding issues with the changeover. The key contribution of this chapter is to integrate the applications of 6G in Industry 5.0 with assimilation and various demanding technologies such as IoT+5G+cloud+AI/ML+robotics and many more. Suggested design and technology roadmap demonstrates the fundamentals of Industry 5.0 and 6G networks. The chapter examines possible situations based on the most common use cases [4, 5]. In order to meet the increasing demands of users in Industry 5.0 and 6G networks, the author's study indicates that convergence and integration of recent communication technologies, network architectures must go hand to hand.

The work in chapter is presented through various sections. Section 2.1 briefs about the state of research in the field of Industry 5.0 and 6G networks. Section 2.2 describes about the present work in the field of Industry 5.0 and 6G. The generations of the Industrial Revolution are covered in Section 2.3. Section 2.4 focuses on components of Industry 5.0. Section 2.5 goes into further detail on Industry 5.0's enabling technologies. The problems and possibilities are covered in Section 2.6. Section 2.7 puts focus on the enhancement in Industry 5.0 and 6G technologies and Section 2.8 ends with concluding remarks.

2.2 Related Work

This section excels the finding and contributions by different researchers in the direction of development of Industry 5.0 along with the use of 6G.

Additionally, it offers suggestions for combining different technologies to achieve Industry 5.0 goals and future directions for new proposals.

Industry 5.0's primary goal is to automate business operations while simultaneously satisfying client demands. It also advances the notion of human–robot collaborations and provides the study of various parameters in collaborative working of human and robots for various applications. Industry 5.0 seeks to improve worker-intelligent production system interaction in order to restore the human element to industry [1]. Humanities and the sustainability of work cultures are crucial factors to take into account in the Industry 5.0 as compared to Industry 4.0. There is a research deficit in the areas of accountability, safety, and sustainability as we go from Industry 4.0 to 5.0 [2, 3]. Industry 5.0 is supported with the advanced technologies and next-generation networks. The 6G network have capabilities to provide multi-terabyte of data per second; to achieve this, it requires an intelligent and autonomous network supported by many advanced technologies. Also, decentralized approach with design issues of propagating new technological evolution is expressed [4, 5]. The concept of using various technologies for the data collection, analysis, and control comprising AI, big data, ML as well as IoT are explored in [6, 7]. The research considered in [8, 9] provides brief discussion about survey of different technologies such as AR/VR, IoT, AI/Ml, cloud computing, collaborative robots, block chain, etc., and is presented in line with 6G communications and development of Industry 5.0. The paper [9] also expresses the functionality and issues of human–robot working together in the production lines. The fundamental knowledge bases required for the socio-economic considerations and development of Industry 5.0 according to business perspective is expressed in [10] and relationship of application as smart city with Industry 5.0 representing smart society in [11]. In [12], the concept of a sustainable smart product is explained by contrasting the roles of Industry 4.0 and Industry 5.0. While Industry 4.0 is known for mass production, Industry 5.0 combines automation and sustainability with the aid of leading-edge technologies to satisfy a range of customer demands. In [13], the Industry 4.0 competency model is put to the test using real-world scenarios. Paper [14] discusses the integrated, ubiquitous, intelligent, and decentralized architecture of 6G based on artificial intelligence. It also focuses on the design issues and enabled technologies in support to AI. The work concluded with proposal of decentralized architecture for 6G. In [15], the idea of human–robot co-working is proposed as collaborator rather than competitor. It also discusses the improvement features used in Industry 5.0 in accordance with technology and manufacturers. The most recent study [16] outlines the Industry 5.0 requirements for IoT and the absolute innovation framework. The review gives brief overview about requirements to maintain energy and cost-efficient 6G network in support of Industry 5.0. The important initiative, key challenges, and major case studies for 6G network applications are also highlighted in the chapter.

2.3 Industrial Revolutions

The transformation of manufacturing industry has been drastically changed with the involvement of digital technologies and beyond. Every new stage in industrial revolution demonstrates how far the manufacturing process has come to alter the relations of humans with business and industry. The section put forward, how the manufacturing sector is integrated with technological evolutions not only for the benefits but also for socio-economic reforms. The complete structure with key elements of every industrial revolution is given in Figure 2.1.

2.3.1 Industry 1.0

The transition to new industrial processes, utilizing water and steam started by replacing the human, termed as First Industrial Revolution and is commenced roughly around 1760. The transformation helped to change the textile and transportation sectors. The revolution gained popularity due to easy working with machines for production of items and availability of fuel like steam and coal. The base of fuel was coal and water, while machines are completely handled by manual process, hence could be termed as "Mechanization."

2.3.2 Industry 2.0

Around 100 years later to Industry 1.0, in 1860, the Second Revolution got underway, primarily in Europe and in the United States. The revolution is termed as "Technological Revolution," since most of the machines were operated on the electrical energy. Even bigger productivity and more complex machines were made possible by the improved electrical technology. Some automation has been introduced in the control of machines like lathe machines. The Second Revolution is specifically based on the "Electrification."

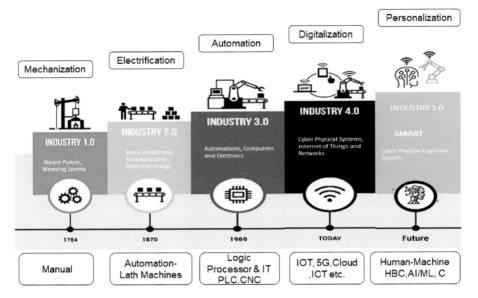

Figure 2.1 Industrial revolutions.

2.3.3 Industry 3.0

It was the era of automation with computer and software technology. Industry 3.0 implemennted after @ 100 years from industry 2.0 i.e 1870 to 1960. The automation in production takes place with integral approach of computer, electronics, and IT (Information Technology). The access of renewable energy, connectivity and Internet access was prime and easily available to increase the production. In Industry 3.0, the programmable logic controllers (PLCs) added the future of automation replacing humans in the manufacturing process. Even though, automation was used in production, human input and intervention were still necessary. Tasks and processes were automated by inputs from human and not by data.

2.3.4 Industry 4.0

In Fourth Industrial Revolution, production and control were handled by intelligent machines having capabilities of storage, communication and cooperation among themselves without the need of human involvement. The Industrial Internet of Things (IIoT) enables this information sharing. To make Industry 4.0 active and functionally operating following are the key elements:

- Cyber–physical system: Computer-based algorithm enhances the performance of a mechanical device.
- Internet of Things (IoT): It increases the monitoring, control and analysis capacity of the interconnected things, such as networks of machine devices embedded for specific tasks with sensing.
- Cloud computing: It provides the virtual back up of the data required for hosting of networks.
- Cognitive and quantum computing: It aids in the creation of solutions using cutting-edge technology like AI/DS.
- The technologies that support for industry 4.0 are IoT, ICT, CC, AI/ML, and Big data with 5G.

2.3.5 Industry 5.0

It is an advancement of Industy 4.0, where customers gets the things what they want ie. Personalization, Terms as Cyber physical cognitive systems (CPCS). Industry 5.0 aims for the makeover of manufacturing industry into SMART IoT-empowered factories utilizing cognitive and computing facilities with interconnected devices through cloud-servers. Collaborative working of machine, human hands and brain will be the main emphasis of Industry 5.0. This indicates that efficiency of production can be increased by finding ways for man and machine to get along and work together to get real-time data from the field. Technologies that support for implementation of Industry 5.0 are, ICT, robotics, AI/ML, human bond communications (HBC), and 6G. Overall, Industry 5.0 development may turn out to be the full realization of what the designers of Industry 4.0 could only have imagined at the start of year 2010s. Connection between computers, robots, and human employees will eventually become more meaningful and mutually illuminating as AI advances and manufacturing robots acquire more human-like characteristics.

2.4 Industry 5.0 Human-centric Approach – Components

Industry 5.0 targets to deliver high-quality, dependable communication, finding ways of man−machine interaction for efficient production and delivery (personalization). According to end-user and network operator requests, the next generation of networks and technological evolutions will include new features including mobile data volumes, connected device numbers, low latency, end-user data rates, channel bandwidth, reliability, connectivity, convergence, cooperation, and content for devices. All of these factors must be taken into consideration to ensure that 6G networks can provide data from flexible service platforms to a fully digitalized and connected world. Figure 2.2 explores the constituents that make up Industry 5.0.

The inner layer has the integration of two important aspects as cyber physical systems (CPS) responsible for providing the technological as well as mechanical platforms. It also provides the way of predicting the data from various sources and does the efficient communication with skilled human operators. The inner layer of system is cognitive with CPPS and human workers. It provides the perception about the data received from environment using sensors, cognition of various technologies used for computation and communications, interacts with various devices and service providers. Outer layer provides the actual information of sensing, technology, computations, communication and services required for human values.

Sensors: It provides the sensing of real-time data used for the working process in production, many type of sensors used for data gathering are motion, health, environment, machine, tactile, audio, visual, etc. (i.e., perception).

Technology: It provides information on how many technologies, including IoT, ICT, cloud computing, AI/ML, AR/VR, big data, robotics, 5G and 6G, are being integrated and transformed to enable Industry 5.0, or cognition.

Algorithms: Various simulations algorithms are used for data computations and AI/ML algorithms are used for the analysis of real-time data in the repetitive manner. They may use the stored datasets or real-time data for predicted results.

Services: In Industry 5.0, COBOTS are working as cyber–physical systems and need the interaction between mobile devices, tele-operated systems, natural language processing and generation unit, digital assistance, social networks, and human–machine interactions.

Human values: It represents the respect for people which comes by education, communication and leadership: Also, Value for Customer comes by personal experience and empowerment.

While doing the operations with Industry 5.0, it is crucial to take the technological conversions into account, network and applications, cooperation with machines and humans, contents and cost which all leads to connectivity.

The fundamental hypothesis is that human civilization is transitioning from its fourth stage, the information society, into its fifth stage, which has been dubbed the imagination society.

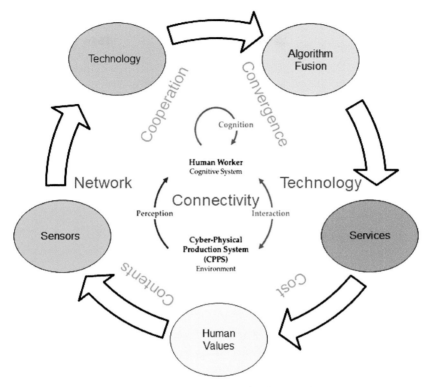

Figure 2.2 Constituents of Industry 5.0.

2.5 Technology Enablers of Industry 5.0

Leading to Industry 5.0, various societies as human (Society 1.0), agricultural society (Society 2.0), industrial society (Society 3.0), information society (Society 4.0), and innovation society (Society 5.0), are gaining attention for mass customization and experience of customers through digital transformations. In, all the revolutions, technology plays important roles, not only to support but functions also. Details of enabling technologies which can be used in Industry 5.0 are shown in Figure 2.3. It consists of industrial block chain, nmWave drones, robotics and automations, IoT, GICT, big data, computing (cloud, quantum, edge, etc.), 5G and beyond; 6G, mixed reality (AR/VR,), AI/ML, etc.

In the enlisted technologies, IoT is used for monitoring and control actions; ICT is used for information communications in compressed and

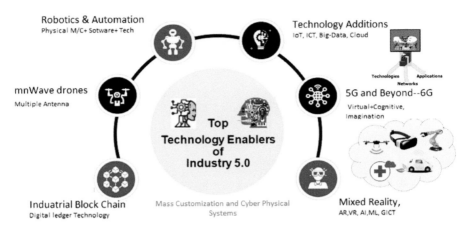

Figure 2.3   Top technology enablers of Industry 5.0.

encoded way; AI and ML algorithms are used for analysis of data; nmWaves are used for the multilayer networks; computing techniques are used for the decision-making capabilities as well as for virtual storage of data. Internet of Things along with ICT support for the theme of 5G and with added computational intelligence using AI/ ML algorithms meet the requirement of 6G. Anyone can find applications that are not limited to telemedicine and health care, smart city, smart homes, smart transportation, industrial IoT changed to PIoT (Industry 5.0), smart agriculture, smart e-commerce, finance, banking, smart automotive (V2X communications), smart robotics and automation, social media, data security, entertainment, gaming, smart education, astronomy, etc., but also in service automations. The certain characteristics of different technologies used in enhancement of Industry 5.0 operations are listed as:

Industrial block chain: Utilized for digital identity creation, operational transparency, and decentralized management

6G and beyond: Knowledge detection, clever resource mapping, fast data speeds, small latency, and ultra-high trustworthiness

COBOTS: Increased productivity, ruggedness, more consistent, reliable, and accurate

Artificial intelligence and ML algorithms: Greater efficiency, more accurate, quality control, instant analysis, and decision-making

Big data analysis: Customization, accurate, precise and fast decision-making, and real-time forecasting

Internet of everything: Monitoring and control, cost reduction, intelligent network operations with things generating variable data, connectivity, intelligence, self-upgradable, and dynamic.

Computing: Storage, increased reliability, accuracy, security, authentications, and extended interoperability.

Mixed reality: Error forecast, customized design, prognostic repairs, training, and instruction.

nmWave drones: Multilayer applications in 5G and 6G, precise analysis, and control.

2.6 Opportunities and Challenges in Industry 5.0

2.6.1 Opportunities of Industry 5.0

The Industry 5.0 provides opportunities in various areas such as:

- **Job opportunities:** In innovation and creative thinking, by means of technology, handling COBOTS, designing AI algorithms and many more.
- **Increasing automation:** In industrial processes gives clients more customized options, maximizing human productivity.
- **Higher-value employment:** skilled human resources are utmost important to integrate latest technology with machine interfaces.
- **In planning:** As human and machines are working together it is important to provide accurate and correct information, it also allows to design functions for personal products.
- **Safety considerations:** Due to collaborative working of machines and humans on same floor, possibility of hazards can not be ignored since machines like COBOTS are working in auto mode.
- **Infrastructure and money:** If appropriate money and infrastructure are available, have strong concerns about the start-ups and business owners in imaginative and creative fields to create new products and services relevant to Industry 5.0.
- **Research and development:** Industry 5.0 places more importance on the subject of human–machine interaction and provides a broader platform for this type of work.
- **Quality service:** Industry 5.0 is capable of offering service at distant localities, for example, in healthcare industry such as medical surgeries in rural areas by robots.

2.6.2 Challenges of Industry 5.0

Industry 5.0 challenges are not only at network, device, or technology level but also at enterprise (end users) level and are shown in Figure 2.4 including:

1. **Infrastructure: flexibility, cost-effective, support:** Industry 5.0 demands for the flexible infrastructure which cost more since it has to be modified according to the needs of customer and match with cutting edge technology requirements. Reliability and stability are the major concerns.

2. **Human−machine collaborations:** The aim of Industry 5.0 is to increase the productivity by collaboratively working with machines (Robots). Care should be taken to avoid the accidents with intervention of machine and human at a time. For maximum performance optimization, Industry 5.0 revolution will observe meaningfully more sophisticated collective communications between humans, machines, processes and systems.

3. **Communication:** Network accessibility, connectivity, and low-power usage.

4. **Device and resource management:** Because production processes are so automated, it is very challenging to improve workers' abilities, such as teaching them how to use front-line technology and altering their social behavior. The work force required will be split into the various categories as qualified, highly skilled, unqualified with low payment may alleviate in the society.

5. **Security and privacy and authentication:** The data requirement of Industry 5.0 is large and must be verified before applying to complete the processes. It is dangerous to provide unauthorized data since human and machine are working together on the same floor. At the same time due to its increased connectivity information should be kept in encrypted form to avoid cyberattacks.

6. **Regulatory mechanisms and business strategy:** Due to higher degrees of automation in the sectors, the Industry 5.0 needs must be satisfied by changing the present company strategies and business structures. Emphasis on customer-centric operations in corporate strategy will grow because of mass personalization. Customer partisanship shifts with phase, which makes it challenging to company strategies and organizational structures to shift frequently. According to client preferences and requirements industrial business strategies are dynamically changing to maintain competition.

Figure 2.4 Challenges of Industry 5.0.

7. **Service automation:** Scalable virtualization, closed-loop assurance, and moving toward zero-touch networks are examples of service automation.

2.7 Integration of Industry 5.0 and 6G Technology Enhancements

Resilience, human-centricity, and sustainability are the three pillars of Industry 5.0. Recent developments and inventions in Industry 5.0, advanced technologies bring people back to the production hub to boost productivity in order to satisfy demand. The advanced technologies such as IoT, 5G, 6G, block chain, robotics with next-generation networks (NGNs) envisioned as a combination of numerous approaches exploited in complimentary advancements. The cognitive cyber−physical systems link real, imagined, and virtual worlds. Integration and convergence of numerous technologies in 6G leads to a completely digitized and linked world.

Figure 2.5 gives the fundamental idea of technologies supporting for the Industry 5.0 and recent developments of 6G technology. The end user in today's digital environment expects a completely digitized and connected world of things. The cornerstone for 6G as indicated in eqn (2.1) is the conversion and integration of smart technologies like IoT, ICT, and cloud.

$$6\,G \approx \text{Virtual} + \text{Imagivation} + \text{Cognitive} \approx \text{IoT} + 5\,G + \text{ Cloud } + A\,I/ \tag{2.1}$$

Low latency of communications between 1000x users and its capacity to send data at high speeds (Tbps), makes 6G suitable for a broad range of applications, including e-Health, smart cities, manufacturing sectors, supply chain management, etc. In all, with visions and available technology support the manufacturing sector becomes SMART. It fulfils the demands of individual customer; hence integration of Industry 5.0 and 6G is expressed in eqn (2.2).

$$\text{SMART} \approx \text{IoT} + \text{ICT} + \text{AI/ML} + \text{Robotics} + \text{Computing} \approx 6G \ \& \ \text{Industry 5.0} \tag{2.2}$$

Following the global deployment of 5G networks, future developments of Industry 5.0, which involves the collabarations of humans and machines, will call for the NGNs and 6G, with exceptionally low latency and AI capabilities. The growth of Industy 5.0 and intelligent communication technology (6G) will have to take support of several elements of connectivity and intelligent, deep, holographic, and ubiquitous networks. In addition to that the future research includes more focus on hyper intelligent networking, quantum computing, block chain, ultimate security and trust. Industry 5.0 is expected to

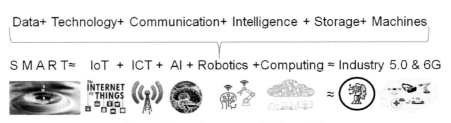

Data+ Technology+ Communication+ Intelligence + Storage+ Machines

S M A R T≈ IoT + ICT + AI + Robotics +Computing ≈ Industry 5.0 & 6G

Figure 2.5   Enhancement of Industry 5.0.

fundamentally alter manufacturing systems and processes by fostering more human–robot collaboration in order to deliver tailored products to clients. The phrase "Industry-5-0-announcing-the-era-of-intelligent-automation with Human and Machine Integration" comes true.

2.8 Conclusion

Industry 5.0 is "Cyber-Physical Cognitive System-Personalization" simply can be narrated as a SMART factory, supporting the human–robot co-working. The AI/ML algorithm, ICT tool, and 6G networks are intelligently controlling it in order to offer at extremely high data rates, capacity, and lowering latency regardless of the number of devices connected to the network at once. From the user's perspective, there will be no decrease in QoS or QoE contrasting to previous revolutions. The new Industry 5.0 revolution provides a path for flexibility and prosperity in the usage of upcoming technologies, which respect people, and values for customers. In summary, for optimal performance optimization, the Fifth Industrial Revolution will witness significantly more classy concerted interactions between humans, machines, processes and systems.

References

[1] Kadir Alpaslan Demir,Gözde Dövena,Bülent Sezen,"Industry 5.0 and Human-Robot Co-working", 3rd World Conference on Technology, Innovation and Entrepreneurship (WOCTINE), Science Direct, Procedia Computer Science 158 (2019) 688–695.

[2] Sandra Grabowska, Sebastian Saniuk, Bożena Gajdzik, "Industry 5.0: improving humanization and sustainability of Industry 4.0", Scientometrics (2022) 127:3117–3144,s://doi.org/10.1007/s11192-022-04370-1.

[3] Francesco Longo, Antonio Padovano, and Steven Umbrello, "Value-Oriented and Ethical Technology Engineering in Industry 5.0: A Human-Centric Perspective for the Design of the Factory of the Future",https://doi.org/10.3390/app10124182,2020.

[4] Nandkumar Kulkarni, Dnyaneshwar S Mantri, Neeli Rashmi Prasad, Pranav M Pawar, Ramjee Prasad, "6G Future Vision: Requirements, Design Issues and Applications", Book Chapter in 6G Enabling Technologies: New Dimensions to Wireless Communication, River

Publications PP: 23-43, Jan 31, 2023 Doi:s://doi.org/10.1201/9781
003360889,

[5] Dnyaneshwar S. Mantri, Pranav M. Pawar, Nandkumar P. Kulkarni
and Neeli R. Prasad, "Ubiquitous Networks: A Need of Future World
of Things", Journal of ICT Standardization, 2021, PP.349-370, Vol 9
Iss 3, 2021, DOI:https://doi.org/10.13052/jicts2245-800X.933Date16/
12/2021

[6] Vural Ozdemir and Nezih Hekim, "Birth of Industry 5.0: Making Sense
of Big Data with Artificial Intelligence, the Internet of Things" and
Next-Generation Technology Policy", OMICS A Journal of Integrative
Biology, Volume 22, Number 1, 2018, DOI: 10.1089/omi.2017.0194.

[7] Dnyaneshwar S Mantri, Pranav M Pawar, Nandkumar Kulkarni, Ramjee
Prasad "Internet of Things (IoT) and Artificial Intelligence for Smart
Communications", Artificial Intelligence, Internet of Things (IoT) and
Smart Materials for Energy Applications, by CRC Press, 12 October
2022, ISBN 9781003220176. https://doi.org/10.1201/9781003220176
Place: Boca Raton

[8] Yang Lua,, Xianrong Zheng, "6G: A survey on technologies, scenarios, challenges, and the related issues", Elsevier Journal of Industrial
Information Integration 19 (2020) 100158.

[9] Amr Adel, "Future of Industry 5.0 in society: human-centric solutions,
challenges and prospective research areas", Journal of Cloud Computing: Advances, Systems and Applications (2022) 11:40https://doi.org/
10.1186/s13677-022-00314-5

[10] Noboru Konno, Carmela Elita Schillaci, "Intellectual capital in Society
5.0 by the lens of the knowledge creation theory " Journal of Intellectual Capital,Vol. 22 No. 3, 2021 pp. 478-505, © Emerald Publishing
Limited,1469-1930,DOI 10.1108/JIC-02-2020-0060

[11] Atsushi Deguchi, "From Smart City to Society 5.0", Society 5.0,https:
//doi.org/10.1007/978-981-15-2989-4_3,2020

[12] Kadir Alpaslan Demir, Halil Cicibaş," Industry 5.0 and a Critique
of Industry 4.0", 4th International Management Information Systems
Conference October 17-20, 2017, İstanbul, Turkey

[13] Schumacher, A., Erol, S., & Sihn, W., "A maturity model for assessing Industry 4.0 readiness and maturity of manufacturing enterprises". Science Direct Procedia Cirp, 52, 161-166., 2016, doi:
10.1016/j.procir.2016.07.040

[14] Xiuquan Qiao , Schahram Dustdar , Yakun Huang, Junliang
Chen, "6G Vision: An AI-Driven Decentralized Network and

Service Architecture", Department: Internet of Things, People, and Processes, IEEE Computer Society, pp33-40, Sept 2020,Doi: 10.1109/MIC.2020.2987738.

[15] Saeid Nahavandi, "Industry 5.0 A Human-Centric Solution", Sustainability 2019, 11, 4371;MDPI doi:10.3390/su11164371

[16] Farhan Aslam,Wang Aimin, Mingze Li and Khaliq Ur Rehman, "Innovation in the Era of IoT and Industry 5.0:Absolute Innovation Management (AIM) Framework" MDPI Journal 2020, 11, 124; doi:10.3390/info11020124

Biographies

Dnyaneshwar S. Mantri graduated in Electronics Engineering from Walchand Institute of Technology, Solapur (MS), India in 1992 and received Master's from Shivaji University in 2006. He has been awarded Ph.D. in Wireless Communication at the Center for TeleInFrastruktur (CTIF), Aalborg University, Denmark. He has teaching experience of 25 years. From 1993 to 2006 he was working as a lecturer in different institutes [MCE Nilanga, MGM Nanded, and STB College of Engg. Tuljapur (MS) India].

Presently he is working as a professor in the Department of Electronics and Telecommunication Engineering, Sinhgad Institute of Technology, Lonavala. He is a recognized masters and Ph.D guide at Savitibai Phule Pune University. He is a Senior Member of IEEE, a Fellow of IETE, and Life Member of ISTE. He has published 06 books, 20 Journal papers in indexed and reputed Journals (Springer, Elsevier, and IEEE, etc.), and 19 papers in IEEE conferences. He is a reviewer of international journals (Wireless Personal Communication, Springer, Elsevier, IEEE, Communication Society, MDPI, etc.) and conferences organized by IEEE. He worked as TPC member for various IEEE conferences and also organized IEEE conferences GCWCN2014 and GCWCN2018. He is a guest editor in STM Journals, Elected as Executive Council member (EC), and vice chair of IETE Pune Local Center (2022-24), His research interests are in Adhoc Networks, Wireless Sensor Networks, Wireless Communications specific focus on energy and bandwidth.

Pranav M. Pawar graduated in Computer Engineering from Dr. Babasaheb Ambedkar Technological University, Maharashtra, India, in 2005, received Master in Computer Engineering from Pune University, in 2007 and received PhD in Wireless Communication from Aalborg University, Denmark in 2016, his Ph.D. thesis received nomination for Best Thesis Award from Aalborg University, Denmark.

Currently he is working as an assistant professor in Dept. of Computer Science, Birla Institute of Technology and Science, Dubai, before to BITS he was a postdoctoral fellow at Bar-Ilan University, Israel during March 2019 to October 2020 in an area of Wireless Communication and Deep Leaning.

He is recipient of outstanding postdoctoral fellowship from Israel Planning and Budgeting Committee. He worked as an associate professor at MIT ADT University, Pune from 2018-2019 and also as an associate professor in the Department of Information Technology, STES's Smt. Kashibai Navale College of Engineering, Pune from 2008-2018. From 2006 to 2007, was working as System Executive in POS-IPC, Pune, India. He received Recognition from Infosys Technologies Ltd. for his contribution in Campus Connect Program and also received different funding for research and attending conferences at the international level. He published more than 40 papers at national and international level. He is IBM DB2 and IBM RAD certified professional and completed NPTEL certification in different subjects. His research interests are Energy efficient MAC for WSN, QoS in WSN, wireless security, green technology, computer architecture, database management system and bioinformatics.

N. P. Kulkarni received Bachelor of Engineering (B.E.) degree from Walchand College of Engineering, Sangli, Maharashtra, (India) in 1996. He has been with Electronica, Pune from 1996 -2000. He worked on retrofits, CNC machines and was also responsible for PLC programming. In 2000, he received the Diploma in Advanced Computing (C-DAC) degree from MET's IIT, Mumbai. In 2002, He became Microsoft Certified Solution Developer (MCSD). He worked as a software developer and system analyst in CITIL, Pune and INTREX India, Mumbai respectively. He has 23 years of experience both in industry and academia. From 2002 onwards he is working as a faculty in Savitribai Phule Pune University, Pune. Since 2007, he was working with SKNCOE, Pune as a faculty in IT Department. Presently working as an associate professor at the School of Computing, MIT ADT University, Pune. He completed a Master of Technology (M.Tech) degree with computer specialization from COEP, Pune (India) in 2007 and Ph.D. from Aarhus University, Denmark in 2019. His area of research is in WSN, VANET, and Cloud Computing. He has published papers in 18 International Journals, 15 International IEEE conferences, 03 National Conferences

Ir. Neeli R. Prasad, CTO of Smart Avatar B.V. Netherlands and Vehicle Avatar Inc. USA, IEEE VTS Board of Governor Elected Member & VP Membership. She is also full professor at the Department of Business Development and Technology (BTech), Aarhus University. Neeli is a cybersecurity, networking and IoT strategist.

She has throughout her career been driving business and technology innovation, from incubation toprototyping to validation and is currently an entrepreneur and consultant in Silicon Valley.

She has made her way up the "waves of secure communication technology by contributing to the most ground breaking and commercial inventions. She has general management, leadership and technology skills, having worked for service providers and technology companies in various key leadership roles. She is the advisory board member for the European Commission H2020 projects. She is also a vice chair and patronage chair of IEEE Communication Society Globecom/ICC Management & Strategy Committee (COMSOC GIMS) and Chair of the Marketing, Strategy and IEEE Staff Liaison Group.

Dr. Neeli Prasad has led global teams of researchers across multiple technical areas and projects in Japan, India, throughout Europe and USA. She has been involved in numerous research and development projects. She also led multiple EU projects such as CRUISE, LIFE 2.0, ASPIRE, etc. as project coordinator and PI. She has played key roles from concept to implementation to standardization. Her strong commitment to operational excellence, innovative approach to business and technological problems and aptitude for partnering cross-functionally across the industry have reshaped and elevated her role as project coordinator making her a preferred partner in multinational and European Commission project consortiums.

She has 4 books on IoT and Wi-Fi, many book chapters, peer-reviewed international journal papers and over 200 international conference papers. Dr. Prasad received her Master's degree in electrical and electronics engineering from Netherland's renowned Delft University of Technology, with a focus on personal mobile and radar communications. She was awarded her Ph.D. degree from Universita' di Roma "Tor Vergata", Italy, on Adaptive Security for Wireless Heterogeneous Networks.

Ramjee Prasad, Fellow IEEE, IET, IETE, and WWRF, is a professor emeritus of Future Technologies for Business Ecosystem Innovation (FT4BI) in the Department of Business Development and Technology, Aarhus University, Herning, Denmark. He is the Founder President of the CTIF Global Capsule (CGC).

He is also the Founder Chairman of the Global ICT Standardization Forum for India, established in 2009. He has been honored by the University of Rome "Tor Vergata," Italy as a distinguished professor of the Department of Clinical Sciences and Translational Medicine on March 15, 2016. He is an honorary professor at the University of Cape Town, South Africa, and the University of KwaZulu-Natal, South Africa, and also an adjunct professor at Birsa Institute of Technology, Sindri, Jharkhand, India. He has received Pravasi Bhartiya Samman Puraskaar (Emigrant Indian Honor Award by the Indian President) on January 10, 2023 in Indore. He is recipient of the prestigious Distinguished Alumni Award under the category 'Excellence in Teaching and Research in Engineering and Technology' for 2023, by the Birla Institute of Technology, Mesra, Ranchi, Jharkhand, India. He has received "Pioneering Visionary Award" by Bihar Jharkhand Association of North America (BAJANA) at New Jersey, USA, on Sunday March 26, 2023.He has received the Ridderkorset of Dannebrogordenen (Knight of the Dannebrog) in 2010 from the Danish Queen for the internationalization of top-class telecommunication research and education. He has received several international awards such as the IEEE Communications Society Wireless Communications Technical Committee Recognition Award in 2003 for making a contribution in the field of "Personal, Wireless and Mobile Systems and Networks," Telenor's Research Award in 2005 for impressive merits, both academic and organizational within the field of wireless and personal communication, 2014 IEEE AESS Outstanding Organizational Leadership Award for: "Organizational

Leadership in developing and globalizing the CTIF (Center for TeleInFrastruktur) Research Network," and so on. He has been the Project Coordinator of several EC projects, namely, MAGNET, MAGNET Beyond, eWALL. He has published more than 50 books, 1000 plus journal and conference publications, more than 15 patents, over 155 Ph.D. Graduates and a larger number of Masters (over 250). Several of his students are today's worldwide telecommunication leaders themselves.

3

Role of 6G, IoT with Integration of AI and ML and Security in Agriculture

**Sunil Kumar Pandey, Abhay Kumar Ray, Smita Kansal,
and Varun Arora**

Institute of Technology and Science, India

Abstract

The convergence of technologies has led to the growth of connected environments and the generation of large amounts of data. However, ensuring the security of these networks has become a significant challenge. Advancements in frequency tuning, antennas, machine learning, AI, and cybersecurity are being pursued to achieve the goals of 6G. Integrating artificial intelligence and machine learning in 6G networks will enhance IoT applications in various domains such as smart cities, smart agriculture, smart transportation, etc.

This chapter focuses on the fundamental concepts, architecture, and applications of 6G, IoT, AI, and ML, emphasizing their integration to establish a secure communication environment. It explores the empowering applications of IoT integrated with AI and machine learning using 6G communication networks. A detailed application of technology illustrates how smart devices and applications can aid farmers in monitoring agricultural fields, optimizing yields, and ensuring profitability through pest control, irrigation management, disease control, and environmental protection. IoT devices can connect to fast 6G networks to access services, receive recommendations, and execute commands from AI and machine learning systems. However, these advancements also pose technological and infrastructural challenges that impact various aspects. The chapter delves into a detailed discussion of each service and technology, concluding with an exploration of future research opportunities and open-ended issues concerning the integration of AI, machine learning, and IoT in future network systems.

Keywords: IoT, AI and ML, Smart Agriculture, 6G, Security and Privacy

3.1 Introduction

3.1.1 Introduction of IoT

IoT (Internet of Things) refers to the interconnection of mechanical and digital machines, physical objects, gadgets, appliances, animals, and humans through standard Internet protocols. This includes not only electronic devices but also non-electronic devices such as water pumps, fans, doors, and more, which are embedded with computational power. The purpose of IoT is to enable these interconnected nodes to exchange data, control operations, and receive commands. By leveraging the power of the Internet, IoT enables seamless communication and integration between various devices and systems, leading to increased automation, efficiency, and connectivity in our daily lives. In other words, Internet of Things is a well-organized system that considers to interconnected physical devices, these devices are accessible and controlled through the web application, standalone software, mobile applications, etc. This application/software is capable enough to access the data from devices, sensors, send commands to actuators and collaborate with cloud applications with the primary goal to connect and exchange data with smart devices, servers, and Cloud system through internet [1]. Though IoT is still growing and in different context, situations, environments, requirements, different variants of IoT are experienced. However, broadly this can be viewed in its 5 designs [2]:

- **CIOT** – Consumer Internet of Things/Commercial Internet of Things
- **IIoT** – Industrial Internet of Things (IIoT)
- **Infra IoT** – Infrastructure Internet of Things
- **IoMT** – Internet of Military Things (IoMT)

Consumer Internet of Things: The term CIoT, or Consumer Internet of Things, generally refers to the utilization of IoT systems and services specifically designed for consumer applications, appliances, and devices. The most commonly used CIoT products are smartphones, smart watches, smart assistants, wearables, home appliances, etc. These devices generally use Wi-Fi, Zigbee, Bluetooth, and short-range communication protocols for the deployment of systems in smaller sites like homes and offices.

The projected global market value for consumer IoT is expected to reach $616.75 billion by 2032, showing significant growth compared to its worth of $221.74 billion in 2022. This growth is estimated to be at a compound annual growth rate (CAGR) of 10.77% from 2023 to 2032 [3].

Commercial Internet of Things: CIoT refers to uses of devices and IoT systems for enterprise and business use. Uses of CIoT can vary broadly by application areas of different sectors. It can be used in healthcare, big offices, buildings, malls, monitoring environmental conditions, facility management in corporate offices, etc.

Industrial Internet of Things: IIoT (Industrial Internet of Things) generally refers to uses of IoT tools and technologies and principles in industrial operations/automation. It involves the use of smart devices and systems in various areas of industry, such as manufacturing, supply chain management, asset tracking, fleet tracking, and more. The goal of IIoT is to enhance operational efficiency, optimize processes, and improve overall productivity in industrial environments. By connecting machines, sensors, and other industrial equipment to the Internet, IIoT enables real-time data monitoring, analysis, and control, leading to improved decision-making, predictive maintenance, and cost reduction. It offers the potential for increased automation, connectivity, and integration across different components of industrial systems, ultimately driving innovation and transformation in the industrial sector. The industrial IoT sector is considered to be the most dynamic and focused segments within the boundaries of applications which uses IoT systems. It focuses to improve productivity, efficiency, predictability, etc. in an industry. The deployment of IIoT services are mostly done in the large-scale industry/factories, manufacturing plants, and assembly lines. The application of IIoT is generally available in automotive industry, logistics, agriculture and ethanol production industry, etc.

Infrastructure Internet of Things: Infrastructure IoT is a subset of the broader Internet of Things (IoT) that focuses on enhancing the reliability, efficiency, cost savings, and maintenance of various infrastructure systems. It involves the application of IoT technologies and principles to critical infrastructure such as transportation networks, utilities, buildings, and other essential facilities. It is basically a subsection of industrial IoT but due to its significant importance in smart infrastructure treated as separate.

Internet of Military Things (IoMT): It generally refers as BIoT (Battle-field IoT) or the Internet of Battle-field Things. IoMT systems use to deploy smart weapons, surveillance system in border or military areas to collect the data for physical sensing, situations awareness, improve response time of different security or attack systems, attack/blast risk assessment etc.

3.1.2 Major components of IoT-based system

IoT systems have six major components which enable the smooth working of the system [4].

- **Sensor actuators and controller:** Sensors are one of the most important components that continuously collect the data from the environment, site, and physical objects. The sensor nodes transmit the sensed data to embedded controller and if the data meets appropriate criteria the controller activates/deactivates appropriate device or set of devices for corrective measures. The device actuation takes place through actuators like Relay, solenoid valve etc. by activating or deactivating a smart device. At last the controller transmits the actuation data to the upper layer of system like local server or cloud system.

The semiconductor industry has made significant advancements in producing micro smart sensors suitable for a wide range of applications. These sensors include but are not limited to light dependent resistors

Callout of Figure 3.1

- **IoT systems consists of internetworked IoT devices with sensors and actuators.**
- **IoT devices can store their data on cloud directly or via local server for further analysis.**
- **IoT systems also have appropriate user interface to access data, report and status**

Figure 3.1 IoT system with local server.

(LDRs), noise sensors, smoke sensors, DHT-11 and DTH-22 temperature and humidity sensors, pressure sensors, light intensity detectors, moisture sensors, as well as RFID tags and readers. These cutting-edge technologies have revolutionized the field of sensing, opening up new possibilities for various industries.

- **Communication protocols:** Everything from ordinary objects to industrial tools have covered as a part of the IoT. Therefore, reliable connectivity and seamless communication required for exchange the data among the devices, for this purpose IoT applications currently uses faster communication protocols like Wi-Fi, Bluetooth , cellular (3G, 4G, 5G), Z-wave, Zigbee, LoRa, etc.

- **Gateway system:** IoT gateway provides the bidirectional data connectivity between sensor nodes/internal system and cloud system using standard communication protocols. It is used to maintain the interpretability among connected divides/nodes and cloud server and it also translates different network protocols data formats. The IoT gateway provides a certain level of security and encryption technique to improve the data security, it behaves like a middle level between the sensor nodes and could provide secure communication and prevent the sensor nodes from unauthorized access and malicious attacks.

- **Local server:** Local server handles the local data storage for the surveillance system; do the local data processing to reduce the overall response time for time-sensitive data processing; selective forward data storage on cloud server because in cloud server storage has significant cost for storage and processing. Local server also receives the request from the user and provides appropriate response to user for resetting a device or accessing the real-time data from sensor nodes or stored data.

- **Cloud server:** The Internet of Things (IoT) has given rise to vast amounts of data generated by devices, sensor nodes, applications, and users. Effectively and efficiently managing this data is crucial. IoT cloud platforms offer a range of tools and applications that facilitate the collection, conversion, processing, management, and real-time storage of large volumes of sensor data. With IoT cloud solutions, authorized users, APIs, and services can access stored data efficiently, whether locally or remotely, via the Internet or standard protocols. This accessibility enables timely critical decision-making as and when needed [6].

IoT cloud contains finely tuned and high performance internetworked servers to carry out data processing with very high speed which came simultaneously from large number of IoT-based devices. It also does the traffic management and provides accurate analytics.

- **User interface:** User interfaces play a crucial role in IoT systems as they are the visible and tangible components that provide a user-friendly experience. It is essential for application developers and designers to prioritize a well-designed user interface to minimize user effort and enhance system accessibility. By creating intuitive interfaces, users can easily operate the IoT system, leading to increased interactions and improved overall user experience.

3.1.3 6G and IoT System

Sixth generation cellular wireless communication system (6G) is the successor cellular technology of 5G cellular network. The research and development for 6G started in 2020 and it is expected to launch commercially in the year 2030 with many services like very low latency (1000× faster response than 5G), and high data rate like 1 TB per second with better location awareness, imaging and presence technology [6]. 6G network services can work together with AI/ML and deep learning to improve decision making processes and 6G computational infrastructure can include services like fast data transmission, data sharing, contextual awareness, better data security and analytics services etc.

IoT with 6G services will provide better services for their recipients with improved data rate, low latency, location awareness, data privacy and security, AI-enabled services, etc., which may provide better use cases in the fields of commercial applications like healthcare system, smart agriculture, smart city, automatic vehicle, facial recognition, law enforcement for better governance and better social credit system in banking area , sensory interfaces which feel like real-life hologram applications, etc. These kinds of applications in real life not only boost the application of IoT but also it might be expected that 6G makes IoT-based applications more reachable to general people.

3.2 Integration of AI (Artificial Intelligence) in IoT Systems

Artificial intelligence involves developing machines that have the ability to make decisions similar to humans. IoT along with AI has shown a new pathway to the technology. The IoT devices generate a large volume and velocity of data which can be used in effective decision-making by using

Callout of Figure 3.2

Data Collected from the IoT Devices are provided to Artificial Intelligence System. Artificial Intelligence System analyzes the data and provides information for better decision making. As shown *in Figure- 3.2*

Figure 3.2 IoT with AI.

AI/ML techniques and algorithms. The root of AI was grounded in the year 1950, when Alan Turing tested whether a computer can reach human intelligence, but the real power is yet to be measured.

According to the future market insight, in the year 2021, the global AI market combined with IoT is currently valued at $73 billion, and it is projected to reach $142.4 billion by the year 2023 [7]. AI can be combined with various other technologies which can find, analyze, and perform from their experiences and help in better decision-making.

3.2.1 Role of AI in the Internet of Things

Devices or objects are interconnected through the Internet, generating and sharing a huge volume of data without human intervention. The data generated by IoT appliances undergoes processing and analysis to enhance the operational performance of these devices. Artificial intelligence plays a vital role in analyzing the data through the utilization of algorithms based on artificial intelligence and machine learning, facilitating valuable insights and improvements [8].

An increase in IoT devices and gluttony for accessing and sharing data and information among these devices results in the consumption of more electricity and a high generation of energy. Artificial intelligence is used in the optimization of electricity consumption and energy production.

Due to the heterogeneity and scalability of networks, security is another concern. There are different areas where attacks can take place, such as on devices, during data transfer, on a cloud, during analysis, etc. To secure the system, artificial intelligence can be used to monitor, threat analysis, trust management, etc.

3.2.2 Role of AI in the realm of 6G communication

The demand for wireless networks is increasing, as services based on the Internet of Everything (IoE) [9] are becoming popular. The progression of networks has advanced from initial-generation (1G) networks to the latest fifth-generation (5G) networks (illustrated in Figure 3.3), with various changes in data rate, latency time, reliability, energy consumption, spectrum, etc. [9].

It is anticipated that 5G can handle the wide range of services needed by the smart Internet of Everything, however, how effectively it will meet the requirement of future smart communities is still unpredictable.

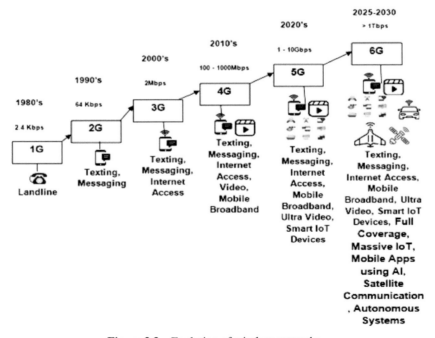

Figure 3.3   Evolution of wireless networks.

Usage of the Internet is increasing rapidly, thus it gives a new edge to IoT applications. IoT applications are not limited to terrestrial but are also available aerial and underwater. Gathering and sharing data/information from various devices with different mediums of communication is the biggest challenge. 6G communication will leverage ultra–large-scale multiple-input multiple-output (MIMO) technology combined with precise channel state information to enable large-scale and high-speed communication. The application of AI technology can enhance the performance of the communication channel, leading to improved efficiency and effectiveness [10].

3.2.3 Machine learning uses in agriculture

As the 5G network is already employed in the world's network, it is expected that soon it will be replaced by the 6G [11]. Recently, vehicle ad hoc networks, agriculture applications, medical applications, etc. have gained the attention of researchers [12]. Recent advances in machine learning have led to solving problems in the wireless domain [13]. Machine learning algorithms allow a computer to learn. Machine learning algorithms will provide basic functionality to ensure the efficiency of future wireless communication networks. As day by day a lot of data traffic is being generated, machine learning algorithms will contribute a lot in the development of six generations of the wireless network. Machine learning is the process that involves learning in computers through observations and experimentation [14]. Various machine learning model and services are available to train predictive models on their data [15]. As machine learning task is associated with many data analysis tasks, it becomes very important for the implementation of a 6G network.

Machine learning algorithms are categorized into the following four categories based on the result to be obtained:

a) Supervised learning: Where the algorithms generate a function that forms a relationship between input and output. For the development of the predictive model, the supervised learning approach uses classification and regression techniques. Classification algorithms are used to predict or categorize the data into various classes. Commonly used classification algorithms are k-nearest neighbor, support vector machine, naïve Bayes algorithm, neural network classifier, and decision tree. Regression techniques are used to predict the continuous data range. Commonly used regression techniques are linear regression, nonlinear regression, logistic regression, stepwise regression, and adaptive neuro-fuzzy learning [16].

b) Unsupervised learning: Where models take only inputs, output values are not available. According to the similarity between the data, clusters are formed. Elements present in one cluster are similar while elements in different clusters are different. The most used clustering algorithms are *K*-means clustering, Fuzzy c-means clustering, hidden Markov model, and hierarchical clustering.

c) Semi-supervised learning: Where models takes input as labeled and unlabeled data to generate the suitable algorithm. Here the model is trained using the labeled data while testing is done on unlabeled data [16]. Algorithms used are Gaussian Mixture Model (GMM), manifold regularization, and semi-supervised SVM.

d) Reinforcement learning: Where the algorithms learn a way where algorithm learns a policy how to behave to map between input and output. Here the agent learns after the interaction with the environment and then observes the results obtained in the interaction. Commonly used algorithms in this category are Q-learning, State Action Reward State Action, and Deep Q-Network.

As the use of Internet and Internet-based applications is increasing in daily life, the future will be of IoT. When the systems will be upgraded to IoT-based systems, a number of issues will also be evolved with it. The domains with IoT are agriculture, medical, manufacturing, etc. Security will be a major concern in this regard. Machine learning techniques are able to provide embedded intelligence in IoT-based devices and networks [17].

As IoT devices share big volume of data, usual methods of data generation, collection, and processing might not be useful. So machine learning algorithms are treated as the best way to provide inbuilt intelligence in such devices [17]. Machine learning algorithms help to extract useful information from devices or human-generated data. Machine learning algorithms are used vastly in many applications that include computer vision, fraud detection, malware detection, etc. Machine learning methods are adopted to solve various challenges in network- based applications like localization, topological changes, fault tolerance, data integrity and energy consumption, etc.

As IoT is playing an important role in current era and a majority of people are getting benefitted with the applications of IoT, farming and agriculture contribute significantly to the GDP of a country. Agriculture industry is vastly using IoT for diagnostics and control [18]. It can help to evaluate various filed variables like soil state, atmospheric conditions, correct requirement of manure, etc. This can also help farmers determine the accurate time to harvest. Machine learning models have proven best in extracting the best value from precise measurement of various soil quality, proper time to harvest and irrigation, to fetch the weather conditions, etc. with automated devices [19].

As agriculture is playing a very important role in the life of every one, a large number of authors have worked in the implementation of machine learning and IoT in agriculture. Authors in [20] proposed a prediction model to identify a disease in the apple orchards in Kashmir Valley using data analytics and machine learning in IoT systems. Authors in [21] comprehensively surveyed applications of machine learning in machine vision and represented its applications in agricultural areas. Authors in [22] presented a work to detect cassava leaf diseases. They used convolutional neural network to achieve high accuracy. Authors in [23] studied machine learning and deep learning approaches in detection of plant diseases in four crops – tomato, rice, potato, and apple.

3.2.4 Security concerns in transition from 5G to 6G

According to a report published by Accenture, cyberattacks against businesses increased by 31% from 2020 to 2021. Organizations experiencing a successful security breach in their businesses, especially in supply chain which has increased at a serious level of 44%−61% as compared to the same time period. This scenario could become more common as countries around the world rolling out the 5G. On the other hand, 6G continues to focus the efforts with 5G to improve network cybersecurity. Though the 6G technology is still evolving and a globally acceptable definition is still being formulated by international standardization bodies, however in different corners of various forums the discussions on how to improve upon the 6G cybersecurity, including optimizing cybersecurity in the Internet of Things (IoT), user data security, and increasing its applicability have been happening.

It is obvious that every network in parallel supports varying number of connected devices within a given geographical area. To understand, it should be noted that a 4G network has capability to connect maximum 2000 concurrent devices per square kilometer (0.38 square miles). However ongoing researches are depicting that this number would increase to 1 million devices connected in the same range with 5G and this is also predicted that this number would even increase more with ambitious 6G networks. This would be supporting up to 10 million devices connected in IoT framework in the same region. It should be noted that the devices increase vulnerability of being probable nodes, known as "attack surfaces" or attack vectors, where unwanted users without authorization can get in to the system to capture data. Thus with increased connectivity, would result in equal proportion of corresponding increase in potential to be the threat of cybersecurity risks. In addition to, the devices such as phones, tablets, smart devices etc., which are participating in the network under use, can be a potential threat irrespective of their vertical and domain whether it is a medical equipment's, industrial machinery or computers. Further, to networked devices, the attack surface does include the infrastructure too. Hence, 5G would eventually move to 6G network, and thus increasing the volume and density of connections to billions of the IoT devices, the RAN (radio access network) that connects users with the cloud, computing resources of homogeneous networks and multivendor environments. This creates a need to make sure that every component of the said network infrastructure must be reliable, available, and safe in cyber environment and supports critical services.

As we march ahead and reach closer to 6G technologies, the idea of "information security by design" is gaining ground for probable and acceptable solution for protecting the ever increasing attack surface. The acceptability of the cybersecurity concept is evolving, transitioning from a reactive model of developing security solutions to a proactive approach. This new perspective emphasizes embedding cybersecurity into 6G-enabled devices right from their inception. Although the concept is new, organizations should consider cybersecurity-by-design thinking to present as a broader perspective of plan of action for addressing the issue of prevention and managing cyber threats.

Emulation is becoming a key component of security planning for a secure environment. The capability for digitally replicating risk incidents empowers a security team to identify and address vulnerabilities and misconfigurations. Additionally, it allows for the practice of cybersecurity strategies within an authentic virtual environment. This allows security teams to gain experience in detecting broader security breaches, identify threats, and take remedial action quickly when threats do surface. A wide range of technologies, including AI, advanced sensors, optics, cloud computing, high-speed digital, satellites and robotics, will rapidly evolve, combine, and expand over the next decade, enabling new usage models enabled by 6G.

6G cybersecurity systems are making security tactics that were once ubiquitous obsolete. This should be remembered that a certificate-based encryption is in fact a cyber-environment. This implies, for instance, that the

6G cybersecurity system verifies whether users have permission to access specific software. This shows that 6G will also be a beneficiary from new existing security approaches. These new approaches to cybersecurity solutions will enable the robust Zero Trust network architecture required for the planned 6G cybersecurity infrastructure. Cybersecurity has been enhanced with the introduction of 5G, but the exponential growth of devices and data promised by 6G requires a robust, real-time cybersecurity response. The emergence and substantial embedding and deployment of artificial intelligence (AI) and machine learning (ML) will be key components of 6G and essential for training cybersecurity systems and algorithms. It also provides an additional layer of complexity for creating a more robust cybersecurity system. But as AI becomes more commonplace, so does the number of bad actors with the skills and incentives to exploit the technology for malicious purposes. Overcoming the vulnerabilities of AI and ML training algorithms are essential for building a resilient, scalable, and secure 6G-powered future.

Cyberattacks can manipulate machine learning models during training or testing, compromising AI predictions. Equally concerning, these attacks can reverse-engineer algorithms to extract sensitive information. A second threat that researchers need to consider is the creation of models designed for malicious purposes. This means committing cybercrime or using it in a military or law enforcement context. The quality of AI algorithms is determined by their reliability, accuracy, and consistency. Tactics to subvert these will have far-reaching implications for all future developments that rely on AI. Therefore, to accelerate innovation across a range of new technologies, AI algorithms must be trained to detect and block adversarial events. The long-term goal of 6G cybersecurity is an autonomous and self-sustaining network that can independently respond to potential threats without disrupting normal use. With these cyber-resilience networks still on the horizon, researchers are exploring adversarial machine learning approaches to train models to identify potential threats and determine appropriate correlated responses. This is the core task of AI in the planned 6G cybersecurity architecture. New technologies always reveal new threats that need to be addressed along with the existing threats that accompany current technologies. Focused on cybercrime prevention, the study highlights the development of 6G cybersecurity solutions that scale to address the threats inherent in the growing multi-vendor market. 6G cybersecurity research is a paradigm shift in how we think about protecting digital data today.

3.2.5 Applications of IoT with AI and ML

- **Smart agriculture:** Farmers today face numerous challenges that traditional farming methods alone cannot address. The evolving agricultural landscape demands innovative approaches to meet the growing requirements of food production and sustainability. Artificial intelligence has brought a new paradigm shift in the field of agriculture, where this new automated method not only fulfill food requirement but also help in increasing employability. This technology provides various information like climate conditions, amount of fertilizers and pesticides to be used, irrigation, weeding, spraying, etc., to farmers which help them in yielding the crop [24].

  To predict climate and weather conditions, first data needs to be collected with different barometrical measurements, then machine learning algorithms are used to refine and different combinations (patterns) are identified and finally artificial intelligence model needs to be constructed on the data discovered by machine learning to generate warnings and information of weather automatically [25].

  Early detection of weeds and pests helps to increase crop production. Artificial intelligence technology like drones and robots can be used to capture images, then get analysis and detection of the presence of weeds and pests can be done using machine learning algorithms [26].

  The smart irrigation system can be constructed using a decision-making machine learning algorithm, which trains the system and collected data can be checked and predictions can be done [27]. More details on smart agriculture are discussed in the next section.

- **Smart homes:** Smart homes (use of IoT) and artificial intelligence have provided a new wing to the technology world. It not only makes the life of human beings easy but also improves the living style.

  The AI-based system can recognize objects and faces easily and inform the owner about the person standing outside the door. In case of any suspicious activity, the notification can be sent to a nearby police station. An automated monitoring system can be used to monitor and estimate security threats using artificial intelligence techniques. AI Algorithms integrated with smoke alarms, help to signal fire and notify the same precisely. AI can be integrated with smart appliances which can be operated with mobile phones.

- **Smart healthcare:** Internet, IoT, and artificial intelligence have shown new pathways to healthcare sectors. Access to quality healthcare is an essential requirement for every individual. The provision of adequate healthcare services is crucial for ensuring the well-being and overall quality of life for people. Automating the healthcare system, where the set of sensors sense the health-related data (like body temperature, heart beats blood pressure, etc.), and AI and ML algorithms are used to do analysis, predict, and notify the health condition of patients to the doctor from time to time. IoT devices are used to gather the data with help of sensors, then data is used for prediction using various AI/ML algorithms. Some illustrations are as given below.

a) Cuff-less blood pressure measurement, pulse pressure waves sensor sense the pulse and multi-instance regression algorithm can be used to check whether the patient is having high or low BP [28].
b) Heart disease prediction can be done using a support vector machine learning algorithm [29].
c) Rate of mental stress can be predicted using a support vector machine learning algorithm [30]

3.3 Smart Agriculture with 6G, AI, and ML

By 2050, the world population may hit the 9.6 billon, a growth of approximately 16% from the current year 2023 and population of livestock including cattle, sheep/goat, pigs, and poultry might also hit 43 billion, a growth of approximately 43% from 30 billion in year 2023. This kind of growth must need more than 60% food to feed for such population growth [31]. There is scope to increase agricultural lands by only 4%. If farmers acquire these lands for agriculture it gives rise to some other environmental side effects. So it is a great challenge for the agriculture sector to increase 50% grain production by 2050 to feed additional 2 billon people and 13 billon livestock [32]. So we need a farming system for the better use of existing agricultural infrastructure with integration of IoT, artificial intelligence and machine learning and fast communication technology with longer range like LoRa, cellular network (5G/6G), etc.

Smart farming or smart agriculture is an emerging area of research and development for the automation of agriculture. The application of IoT-based solution which uses a set of different sensors interfacing with embedded controllers [33]. IoT-enabled nodes or machines deployed over the farms or agriculture lands generate large amount of data per day related to the plants, soil, environment, disease-related information, yield-related data, etc. These

Callout of Figure 3.4

Services of Smart Agriculture

- **Smart Spryer**
- **Crop and Soil Monitoring**
- **Crop Yield Predication**
- **Predictive Insights**
- **Disease Diagnosis**
- **Agriculture robots**
- **All above services not only boost the yield of crop but also provide precession for the soil and diseases monitoring by using robots as given in Figure 3.4.**

Figure 3.4   Services of smart agriculture.

data can be used with artificial intelligence and machine learning techniques along with the fast communication system like cellular network (4G/5G/6G) to obtain useful insights/analytics related to pest and diseases infection, crop yield forecasting , price predictions using analytics. These kind of insights can help the farmers, governing bodies to manage demand and supply, to decide crop pattern in agriculture field, pesticide usage, etc.

Smart agriculture has some major services as given below:

a) Crop and soil monitoring
b) Smart sprayer/spraying machine
c) Agricultural robots
d) Crop yield and price prediction
e) Disease monitoring
f) Predictive insights

- **Crop and soil monitoring:** Crop monitoring refers to acquiring the information about crop's plant-related information like trunk, leaf, flower and grain diameter, plant temperature, leaf and flow-bud temperature and soil monitoring (measure soil's moisture and temperature), EC (electricity conductivity), pH value, measure the quantity of three big macronutrients of plants, NPK (nitrogen, phosphorus, and potassium), soil water potential, sun light intensity, photosynthetic radiation, volumetric water content, and oxygen level in soil using IoT and supporting tools and techniques which empowered the farmers and crop producer companies to maximize yield, reduce plants diseases and optimize resource consumption [34]. Data collected from the IoT sensor nodes are transmitted back to AI- and ML-enabled cloud-based system for the for analysis, visualization, actuation, and trend analysis.
- **Smart sprayer/spraying machine:** A smart spraying system having cutting-edge technology support and an advanced form of connectional spraying machine which disperses pest control chemicals on the crop more precisely using a set of rule and regulations and make sure that the chemical effect is efficient and sustainable.

  It reduces the use of pesticide by more than 50% which reduces the side effects of pesticides on crop consumer health. It also reduces the airborne spray drift by 87% much makes plants healthier and reduces the droplet size of pesticides. In spite of all above benefits it provides the same pest control in comparison to the conventional pest control system. This system sprays the chemicals where it is needed. It comprises many components of which the most important are:

  a) **Sensing function:** It is a set of sensors which measure the differences in crop, soil, weed, patches pattern using some imaging tools and techniques which make a basis for the spot treatment, variable rate application treatment, or patch treatment [35].

  b) **Response function:** It determines what and where to be done, means whether the chemicals to be spread or treatment to be applied in which area of the plant.

  c) **Spray technology:** It responds to the demand of the treatment for variable rate or spot spray techniques.

The spraying system has an image acquisition camera having the capability to merge frames in 1024×256 size image and transmitted to the centralized server for the real-time processing for the target detection, which uses neural network and deep learning algorithms like CNN (convolutional neural network) for pest and spot recognition and classification after the target detection then weed position and nozzle trigger take place for the proper treatment of plant. This kind of application need fast and very efficient technology like 6G for the image processing application in real time with very less latency time for the actuation of nozzle on a precise location.

- **Agricultural robots:** One of the major benefits of robotics system is their flexibility to perform a variety of odd or common task in any environment. Robots perform tasks more accurately, precisely, and consistently than conventional labor or skilled workers.
Robots can increase crop production and profit margin in agriculture, as they can finish tasks more quickly. Robots can work nonstop (except scheduled maintenance) for 24 hours a day and they do not require breaks, leaves, and sick days. Robots can also do fewer mistakes than workers and save time. Robots can resolve the agricultural labor problems for those countries where labor is very costly.
All these attributes of robots make them perfect choice for agriculture, especially for large farms in size. Traditionally, agriculture requires significant hard labor with considerable harvesting in large areas of crop fields, and processes like planting, irrigation, fertilization, plant monitoring take more time. But in today's scenario, a single robot can perform all these agricultural tasks and reduce dependency on human labor.
As report published by Emergen Research, the worldwide agricultural robots market size is expected to hit 37.24 billion in the year 2027 with a CAGR of 34.4% robust revenue growth [36]. Many robotics companies like ecoRobotix, Agrobot, Blue River Technology, Harvest CROO Robotics are not only investing huge amounts but also doing research and development to make robots more efficient. These companies use technologies like 3D sensing, machine learning, computer vision, neural networks, sensor fusion, etc., to make robots more capable of doing most of the harvesting and other agricultural works more accurately and precisely.

- **Crop yield and price prediction:** The biggest worry for the farmers is the price fluctuation of the crops especially vegetable and fruits. Due to unsteady prices, farmers are most of the time unable to plan a specific production pattern. This kind of problem is highly common in crops like tomatoes, beans, peas, etc., as these vegetables have inherently short shelf time. Farming companies are generally using satellite imagery or quadcopters with image capturing system to capture and monitor the crop's health conditions and perform appropriate actions like spot selection, pesticide selection and spray, on real-time basis. With the help of modern technologies like AI, ML, and big data analytics which use past yield and price data, crop health data v/s current crop data, companies become capable of forecasting the crop yield and price, which provides the important information to farmers and government for better planning related to future price pattern, time for a type of crop to show the maximum profit.

- **Disease monitoring and predictive insights:** Plant's images patterns, heath-related data like diameter of stem, buds, grain, etc., collected by the imaging system and sensor nodes using SD-5P or SD 6P. These both systems can be mounted on the quadcopters to collect the data. The data received from different sensor nodes of the crop fields are uploaded to cloud servers. The cloud servers apply the AI and ML techniques and data analytics, which can help the farmer to provide prior information for disease, classification of plants for weed identification, monitoring the infected plants and taking corrective measures after the disease identification. Smart farming can use data analytics and supporting tools like big data to provide predictive insights for the right time to saw the right plant seed, soil conditions, plant growth pattern, right action according to current weather conditions or upcoming weather conditions.

Many startups are working on AI with integration of IoT in the field of agriculture. Startup3, a Berlin-based company developed a plant disease detector and pest diagnostic application, which takes multiple images of plants or its different sections and detect diseases based on AI and ML to resolve the puzzle of plant diseases.

When we think of such kind of revolutionary technology for modern farming, a fast communication medium like 6G is required along with IoT and AI and ML. 6G provides very fast data transmission upto 1 TBps for better image/video stream uploading which reduces the response time and latency significantly. The feature of 6G technology like location awareness, contextual awareness, presence technology better data security for data in transmission etc. These qualities of 6G will definitely boost the capability of smart farming applications.

6G, IoT, AI, and ML definitely solve the scarcity of labor and resources in agriculture up to a large extent and the combination of these helps the service developers to make powerful application to mitigate the challenges and complexity of modern agriculture. So many big companies are investing and consistently doing research and development in this field. But the big question is, can these combinations of technology replace the knowledge of farmers that they had? The answer is no as of now. But these technologies will complement and provide better support to the farmers to improve farming practices which leads to better yields, improved agricultural practices, better resource utilization, more profits, and better lives of farmers.

3.4 Future Scope

As agriculture is the major occupation of the people and is the basic need for the survival of the people of a country, it becomes very important to precisely determine the various parameters. Although a large amount of work is already being done for the improvement and precisely determining the parameters that affect the quality of agriculture and farming, more accurate and timely decisions are required for improvement in the quality of agriculture and farming. From the literature, it is found that there are limited machine learning-based automated systems that can be used for the timely detection of diseases. Machine learning is poised to revolutionize various aspects of agriculture in the near future. Tasks such as weed detection, plant disease detection, estimation, plant water requirement, and soil analysis are expected to become routine work as machine learning algorithms and techniques are increasingly employed in agricultural practices. This technology has the potential to greatly enhance efficiency, accuracy, and productivity in the field of agriculture. The work may be extended in future by including more accurate data filtration that is received from sensors and to develop machine learning models that will help the farmers to get correct, precise, and timely information so that they can implement the things easily and can work for the growth of their crops.

3.5 Conclusion

In the present scenario, new technology is immersed by combining already existing ones. The scope of these techniques is not limited to specific areas or sectors. The agricultural sector stands as a prime example where technologies such as IoT, 6G, AI, and ML can significantly contribute to enhancing the well-being of farmers. Forecasting of farming external factors data such as climate conditions, soil quality, and pest attacks can be done precisely. It is good to start up with the preplanned schedule at the harvesting time, but external parameters are unpredictable. To develop a system, a large amount of data for training the machine is needed then only it can predict or forecast. A variety of crop-related data are obtained once a year when grains grow, thus databases take time to establish, and thus AI/ML models take plenty of time to be constructed.

The future farming industry will rely on adaptive cognitive technologies and its solution. Though many applications are already available, there is still vast scope for research to work on. Farming still has insufficient services and realistic challenges and demands are untouched areas.

References

[1] N. Ahmed, D. De and I. Hussain, "Internet of Things (IoT) for Smart Precision Agriculture and Farming in Rural Areas," in *IEEE Internet of Things Journal*, vol. 5, no. 6, pp. 4890-4899, Dec. 2018, doi:10.1109/JI OT.2018.2879579.

[2] "Internet of Things: The Five Types of IoT" Web link :-https://syntegra .net/internet-of-things-the-five-types-of-iot/AvailablefromAug.2022la stseenFeb2023

[3] "Precedence Research "Consumer IoT Market Size to Hit Around USD 616.75 Bn by 2032" URL:https://www.globenewswire. com/en/news-release/2023/01/25/2595350/0/en/Consum er-IoT-Market-Size-to-Hit-Around-USD-616-75-Bn-by 2032.html#:~:text=The%20global%20consumer%20IoT%20market, report%20study%20by%20Precedence%20Research.,January 25, 2023 Last seen Feb 2023"

[4] Mr. Rajeev "What are the major components of Internet of Things" Web URL :https://www.rfpage.com/what-are-the-major-components-of-inte rnet-of-things/dataofpublication:31july2022lastseenFeb2022

[5] Maryam Hassanlou, Yi Sun, Fang Fang "Minimizing Response Time of IoT-Based Traffic Information System Through A Decentralized Server System" Journal of Supply Chain and Operations Management, Volume 18, Number 1, March 2020

[6] Emilio Calvanese Strinati , Jose Luis Gonzalez-Jimenez, Nicolas Cassiau "6G: The Next Frontier " arXiv:1901.03239v2 [cs.NI] 16 May 2019 URL :https://arxiv.org/pdf/1901.03239.pdf

[7] "AI in IoT Market" URL :https://www.futuremarketinsights.com/repor ts/ai-in-iot-marketlastseenFeb2023

[8] Mohammad Riyaz Belgaum, Zainab Alansari, Shahrulniza Musa, Muhammad Mansoor Alam, M. S. Mazliham "Role of artificial intelligence in cloud computing, IoT and SDN: Reliability and scalability issues" International Journal of Electrical and Computer Engineering (IJECE) Vol. 11, No. 5, October 2021, pp. 4458~4470 ISSN: 2088-8708, DOI:10.11591/ijece.v11i5.pp4458-4470

[9] Helin Yang, Student Member, IEEE, Arokiaswami Alphones, Senior Member, IEEE, Zehui Xiong, Dusit Niyato, Fellow, IEEE, Jun Zhao, Member, IEEE, and Kaishun Wu, Senior Member, IEEE "Artificial Intelligence-Enabled Intelligent 6G Networks" Volume: 34, Issue: 6, November/December 2020, DOI:https://doi.org/10.1109/MNET.011 .2000195

[10] Md. ShahjalalWoojun KimWaqas KhalidSeokjae MoonMurad KhanShuZhi LiuSuhyeon LimEunjin KimDeok-Won YunJoohyun LeeWonCheol LeeSeung-Hoon HwangDongkyun KimJang-Won LeeHeejung YuYoungchul SungYeong Min Jang "Enabling technologies for AI empowered 6G massive radio access networks" DOI:https://doi.org/ 10.1016/j.icte.2022.07.002

[11] F. Tang, Y. Kawamoto, N. Kato, and J. Liu, "Future Intelligent and Secure Vehicular Network Toward 6G: Machine-Learning Approaches," Proc. IEEE, vol. 108, no. 2, pp. 292–307, 2020, doi:10.1109/JPROC.20 19.2954595.

[12] F. Yang, S. Wang, J. Li, Z. Liu, and Q. Sun, "An overview of Internet of Vehicles," China Commun., vol. 11, no. 10, pp. 1–15, 2014, doi:10.110 9/CC.2014.6969789.

[13] S. Ali et al., "6G White Paper on Machine Learning in Wireless Communication Networks," pp. 1–29, 2020, [Online]. Available:http: //arxiv.org/abs/2004.13875.

[14] J. G. Carbonell and T. M. Mitchell, AN OVERVIEW OF MACHINE LEARNING. Morgan Kaufmann.

[15] C. Song, T. Ristenpart, and C. Tech, "Machine Learning Models that Remember Too Much," pp. 587–601, 2017.

[16] S. Messaoud, A. Bradai, S. H. R. Bukhari, P. T. A. Quang, O. Ben Ahmed, and M. Atri, "A survey on machine learning in Internet of Things: Algorithms, strategies, and applications," Internet of Things (Netherlands), vol. 12, p. 100314, 2020, doi: 10.1016/j.iot.2020.100314.

[17] F. Hussain, R. Hussain, S. A. Hassan, and E. Hossain, "Machine Learning in IoT Security: Current Solutions and Future Challenges," no. c, pp. 1–38, 2020, doi:10.1109/COMST.2020.2986444.

[18] J. Muangprathub, N. Boonnam, S. Kajornkasirat, N. Lekbangpong, A. Wanichsombat, and P. Nillaor, "IoT and agriculture data analysis for smart farm," Comput. Electron. Agric., vol. 156, no. November 2018, pp. 467–474, 2019, doi:10.1016/j.compag.2018.12.011.

[19] A. Vij, S. Vijendra, A. Jain, S. Bajaj, A. Bassi, and A. Sharma, "IoT and Machine Learning Approaches for Automation of Farm Irrigation System," in Procedia Computer Science, 2020, vol. 167, pp. 1250–1257, doi:10.1016/j.procs.2020.03.440.

[20] R. Akhter and S. A. Sofi, "Precision agriculture using IoT data analytics and machine learning," J. King Saud Univ. - Comput. Inf. Sci., vol. 34, no. 8, pp. 5602–5618, 2022, doi:10.1016/j.jksuci.2021.05.013.

[21] T. U. Rehman, M. S. Mahmud, Y. K. Chang, J. Jin, and J. Shin, "Current and future applications of statistical machine learning algorithms for agricultural machine vision systems," Comput. Electron. Agric., vol. 156, no. December 2018, pp. 585–605, 2019, doi:10.1016/j.compag.2018.12.006.

[22] G. Sambasivam and G. D. Opiyo, "A predictive machine learning application in agriculture: Cassava disease detection and classification with imbalanced dataset using convolutional neural networks," Egypt. Informatics J., vol. 22, no. 1, pp. 27–34, 2021, doi:10.1016/j.eij.2020.02.007.

[23] J. A. Wani, S. Sharma, M. Muzamil, S. Ahmed, S. Sharma, and S. Singh, Machine Learning and Deep Learning Based Computational Techniques in Automatic Agricultural Diseases Detection: Methodologies, Applications, and Challenges, vol. 29, no. 1. Springer Netherlands, 2022.

[24] Tanha Talaviya a , Dhara Shah a , Nivedita Patel b , Hiteshri Yagnik c , Manan Shah d "Implementation of artificial intelligence in agriculture for optimisation of irrigation and application of pesticides and herbicides"http://www.keaipublishing.com/en/journals/artificialintelligence-in-agriculture/

[25] Chris Huntingford , Elizabeth S Jeffers , Michael B Bonsall , Hannah M Christensen , Thomas Lees and Hui Yang "Machine learning and artificial intelligence to aid climate change research and preparedness" DOI:https://iopscience.iop.org/article/10.1088/1748-9326/ab4e55

[26] Nidhi Gupta, Bharat Gupta, Kalpdrum Passi and Chakresh Kumar Jain "Applications of Artificial Intelligence Based Technologies in Weed and Pest Detection" Journal of Computer Sciences

[27] Anas H. Blasi, Mohammad A. Abbadi, Rufaydah Al-Huweimel "Machine Learning Approach for Automatic Irrigation System in Southern Jordan Valley" Engineering, Technology & Applied Science Research Vol. X, No. X, 20XX, pp DOI: :https://www.researchgate.net /publication/338169883

[28] F. Miao, Z.-D. Liu, J.-K. Liu, B. Wen, Q.-Y. He, and Y. Li, "Multisensor fusion approach for cuff-less blood pressure measurement," IEEE J. Biomed. Health Informat., vol. 24, no. 1, pp. 79–91, Jan. 2020.

[29] M. Muzammal, R. Talat, A. H. Sodhro, and S. Pirbhulal, "A multi-sensor data fusion enabled ensemble approach for medical data from body sensor networks," Inf. Fusion, vol. 53, pp. 155–164, Jan. 2020.

[30] F. Al-Shargie, "Fusion of fNIRS and EEG signals: Mental stress study," engrXiv, vol. 2019, pp. 1–5, Apr. 2019, doi: 10.31224/osf.io/kaqew.

[31] Livestock and Livestock products by 2050" URL:https://www.research gate.net/publication/344188926lastseen:Feb.2023

[32] How to Feed the World in 2050" URL:https://www.fao.org/fileadmin/t emplates/wsfs/docs/expert_paper/How_to_Feed_the_World_in_2050. pdflastseen:Feb2023

[33] S. R. Prathibha, A. Hongal and M. P. Jyothi, "IOT Based Monitoring System in Smart Agriculture," *2017 International Conference on Recent Advances in Electronics and Communication Technology (ICRAECT)*, Bangalore, India, 2017, pp. 81-84, doi:10.1109/ICRAECT.2017.52.

[34] R. Singh, S. Srivastava and R. Mishra, "AI and IoT Based Monitoring System for Increasing the Yield in Crop Production," *2020 International Conference on Electrical and Electronics Engineering (ICE3)*, Gorakhpur, India, 2020, pp. 301-305, doi:10.1109/ICE348803.2020.9 122894.

[35] "Smart Spraying Technology in Agriculture" URL :https://euristiq.com /smart-spraying-technology-in-agriculture/lastseen:Feb2023.

[36] "World's Top 6 Companies in the Agricultural Robots Market" URL :https://www.emergenresearch.com/blog/worlds-top-6-companies-in-t he-agricultural-robots-marketlastseenFeb2023''

Biographies

 **Sunil Kumar Pandey** is currently working as professor in Institute of Technology and Science with an experience of over 24+ years in industry and academia and having interest in cloud, blockchain, database technologies, and soft computing. Dr. Pandey has been credited with 65+ research papers (including SCI/ Scopus Indexed), three book chapters, and three books with reputed publishers including Springer, IGI, IEEE Xplore, River Press – Denmark, Wiley, Hindawi, and journals/conferences.

He has been a regular author of articles in different print and online platforms including interviews, views and has published 11 edited volumes on different relevant themes of information technologies. He has been providing and coordinating training and consultancy to various reputed organizations including Indian Air Force and has conducted 25+ national/international conferences/summits/conclaves in association with AICTE, CSI, DST, and other leading organizations. He has also conducted large number of FDP/entrepreneurship programs supported by AICTE/DST/UGC/EDI, etc.

 **Abhay Kumar Ray** is presently working as an assistant professor in Institute of Technology and Science, Mohan Nagar, Ghaziabad. He holds degrees of MCA (Master of Computer Application), MBA (Specialization in Information Systems). He is pursuing Ph.D. from SRM University, Delhi NCR Campus, Modinagar, Ghaziabad , India.

He has published several papers in national and international journals and conferences and conducted 25+ workshops on cutting edge technologies in different institutes of India. Mr. Ray has experience of 16 years in both industry and academia. His area of interest is Internet of Things (IoT), web programming, security system and artificial intelligence, and machine learning.

Smita Kansal is presently employed as an assistant professor in the MCA department at the Institute of Technology and Science, located in Mohan Nagar, Ghaziabad. She holds a master's degree in computer applications from Rashtrasant Tukadoji Maharaj Nagpur University. With a combined experience of 14 years in both industry and academia, her primary focus lies in the field of artificial intelligence and machine learning.

Varun Arora is currently working as an assistant professor at MCA department in Institute of Technology and Science, Mohan Nagar, Ghaziabad. He holds M.Sc. (Physics), MCA, and M.Tech in Computer Science. He is pursuing Ph.D. from Jaypee Institute of Information Technology, Noida. He has published several papers in national and international journals and conferences. He has a total 14 years of teaching experience. His area of interest is machine learning and nature inspired algorithms.

4

Visible Light Communications for 6G: Motivation, Configurations and New Materials

Ana García Armada, Máximo Morales-Céspedes, and Ahmed Gaafar Ahmed Al-Sakkaf

Universidad Carlos III de Madrid (UC3M), Spain

4.1 Introduction

Optical wireless communication (OWC) is a recent technology that includes infrared, visible light, and ultraviolet spectrums. Visible light is a small portion of the spectrum, banded between 380 nm (790 THz) and 750 nm (430 THz), which can be detected by the human eyes, as illustrated in Figure 4.1. It increases to 1400 nm (215 THz) considering the near infrared spectrum. Visible light communications (VLC) have a dual purpose; the light-emitting diode (LED) provides illumination and data transmission. Whereas a VLC receiver can be any device that can convert the power of the optical signal into electrical current such as a photodiode, camera sensor or solar cells. It is worth noticing that the receiver size is considerably greater than the signal wavelength. As a consequence, the VLC are not subject to small-scale channel effects such as Rayleigh/Rice fading or Doppler effects. Moreover, coherent modulation, i.e., modulating both amplitude and phase cannot be implemented. Intensity modulation (IM) is considered for data transmission and the signal power is directly detected (DD), which it typically referred to as IM/DD [9]. Finally, notice that the variatons given by the data modulation of the optical signal must be faster than 200 Hz in order to avoid flickering by human eyes [21].

The first known application of VLC is the photophone developed by Alexander Graham Bellfor transmitting voice over sunlight beams. Then, the new era of VLC started in the early 2000s at Keio University Japan,

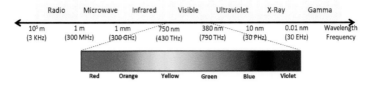

Figure 4.1 Visible light spectrum.

when white light emitting-diodes (LEDs) is used for illumination and communication [14]. Since then, a lot of researchers focus on VLC technology. In this sense, the main advantage of VLC in comparison with RF systems is the wide and unlicensed available spectrum, which attracts many researchers and industries. Moreover, VLC networks are extremely useful for moving data traffic from the overwhelmed RF spectrum to the optical domain, which does not interfere with other systems. In this sense, VLC have been proposed for 6G to achieve key performance indicators (KPIs) such as data rate density about 1-10 $Gb/sec/m^2$ or latency below 0.1 msec [11]. VLC are also suitable for providing connectivity in those environments where RF transmission is inefficient or it is even banned such as coal mines, hospitals, and oil & gas sectors. Besides, VLC is a more secure technology since light cannot penetrate walls or other opaque objects. Focusing on the operational and capital expenditures, VLC can be considered a low-cost technology in comparison with the latest RF devices, comprising mmWaves, since they avoid the need for elements such as up/down converters or high-bandwidth analog-to-digital converters.

Despite all these advantages, VLC still face some challenges. First, line of sight (LoS) between the pair transmitter-receiver is required. Beyond the characteristics of the environments, which may lead to blocking and shadowing effects, the orientation of the receiver can also vary with the human movements. Therefore, the concept of field of view (FoV) does not depend exclusively on the photodiode and lens, it also has a variable component. Moreover, the lack of small-scale effects generates correlated channels responses more likely than in RF systems.

Interference may appear in VLC caused by any natural light source such as sunlight or artificial ambient light, which is mainly treated as noise. However, most of the interference is generated a the neighboring optical access points (APs) that overlap their coverage footprints, which is referred to as intercell interference. Indeed, the presence of intercell interference may increase the noise and lead to degrading the performance of the system, so that interference mitigation techniques are required in VLC. Furthermore, there are still some issues related to the VLC uplink. It is impractical to apply a VLC uplink system in portable devices such as smart phones, laptops, even though some of these devices are provided with camera flashlight or notification indicators because of they are running in low power battery and second it can cause visual disturbance to the user while using the devices.

As commented above, VLC play a major role for future 6G wireless communications. In this sense, well-known concepts for RF systems such as (massive) MIMO, analog/digital precoding or beam selection can be applied to VLC. However, the optical transmission is subject to different constraints and characteristics in comparison with RF. In this chapter, we analyze the elements that compose a VLC system, focussino on alternative configurations and architectures that allow us to maximize the performance terms of capacity and latency of VLC in indoor and vehicular scenarios [4, 8].

4.2 Visible Light Communications System

The basic VLC system is composed of a LED, a photodiode, the drivers that modulate the information, and the microcontrollers that manage the transmitted/received information. In comparison with RF systems, it is worth remarking that VLC employs IM/DD, i.e., the data is modulated varying the input current of the LED from 0 Hz to the maximum modulation frequency. As a consequence, VLC systems do not require complex signal processing chains based on up/down converting the baseband signal to intermediate frequencies. In the following, we describe each of the elements that compose a VLC system.

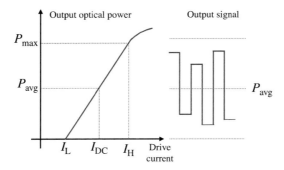

Figure 4.2 Impulse modulation. The input current of the LED is modulated according to the data source signal.

4.2.1 Transmitter

The transmitter is composed of a processor, chipset or microcontroller that generates the data to transmit. However, this signal is typically limited to a low-voltage and current, and therefore, cannot switch on an illumination LED. Therefore, it is necessary to design a driver between the data source and the LED. Notice that data corresponds either to monocarrier modulation, e.g., pulse amplitude modulation (PAM) or pulse position modulation (PPM), or multicarrier modulation, e.g., direct-current offset OFDM (DCO-OFDM) or asymmetrically clipped OFDM (ACO-OFDM) [9]. After that, the signal is transmitted using IM/DD. That is, it is necessary to modulate the drive current of a LED. The concept of impulse modulation is described in Figure 4.2, the current value I_{DC} fixes the drive current that generates the desired illumination, which corresponds to the average signal value denoted by P_{avg}. Notice that the data signal is contained within the range $\{-A, A\}$, so that the average illumination is given exclusively by the polarization current I_{DC}. It is worth remarking that the transmitted signal, which corresponds to the input current that feeds the LED, must correspond to a real and positive value.

The simplest transmit driver is given by a transistor that generates the drive current that satisfies the illumination requirements while the data source modulates that current as it is shown in Figure 4.3. The drive current is inversely proportional to the resistance at the collector of the transistor, i.e., $I_{\mathrm{drive}} \approx \frac{V_{cc}}{R_c}$, while the amplitude of the modulated signal can be set with the resistance at the base of the transistor. Besides, most of the data sources or microcontrollers are not protected to current overloads, which may appear when the LED reflects some of the current injected by the drives. Thus, it is recommended to add an operational amplifier in buffer configuration between data source and the elements of the drivers.

It is worth noticing that the transmission bandwidth is given by the electronic components of the driver and the LED. In this sense, components such as operational amplifiers, transistors, etc., are not typically the main bandwidth limitation. In fact, the cost of these elements is proportional to the maximum modulation bandwidth. However, the main modulation bandwidth bottleneck is the LED. Achieving white illumination require to add a phosphor layer that converts the blue light originally emitted by a LED into white, which involves a bandwidth limitation. In this sense, blue LEDs without the aforementioned phosphor layer provides larger modulation bandwidths. Recently, organic LEDs have been also proposed for achieving modulation bandwidths above 20 MHz. On the other hand, the use of commercial LEDSs

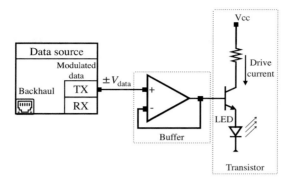

Figure 4.3 Simple transmitter architecture for VLC. The signal from the data source is modulated through the drive current of the LED.

preserving the energy-efficiency has been recently pointed out as a requirement for the success of the VLC technology [7]. However, the modulation bandwidth of this commercial LEDs is typically unknown. In general, the modulation bandwidth is inversely proportional to the illumination capacity of the LED, being necessary alternative measurement methodologies for determining the frequency response of the LEDs. Recentlyt, the laser diodes (LD) have been proposed as an alternative VLC transmitter that obtains higher performance in comparison with conventional LEDs. This makes it a promising technology in the 6G of wireless communications [28]. The modulation bandwidth of LD is much faster than normal LEDs, which can reach up to 100Gbps, satisfying the requirements of the ultra-high data density services in 6G.

4.2.2 Optical channel

The VLC channel corresponds to the sum of a flat fading channel pluse a diffuse component depending on the reflections and the propagation environment. Specifically, the VLC channel at time t can be written as,

$$\eta(t) = \eta_{\text{LoS}}\delta(t) + \eta_{\text{diff}}(t - \Delta t) \tag{4.1}$$

where η_{LoS} is the Line-of-Sight (LoS) component, η_{diff} is the contribution due to the NLoS rays that generate a diffuse component and Δt corresponds to the delay between both the LoS and NLoS component.

The LoS contribution is given by the Lambertian model, which mainly depends on the geometry of the scenarios as can be seen in Figure 4.4. That

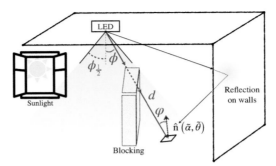

Figure 4.4   Propagation for VLC scenarios. The Lambertian channel model (LoS) parameters are highlighted in blue.

is, the LoS contributions mainly depends on the distance between transmitter and receiver, which is denoted by d, and the irradiance and incidence angles are denoted by ϕ and φ, respectively. Assuming that the incidence angle is within the field of view (FoV) of the photodiode, the LoS component is given by,

$$\eta_{\text{LoS}}(d, \phi, \varphi) = \frac{\gamma A}{d_{kt}^2} R_0(\phi)\, T(\varphi) g(\varphi) \cos^r(\varphi) \qquad (4.2)$$

where γ is the responsivity of the photodiode, A is area of detection and R_0 is the Lambertian radiation intensity, which is given by

$$R_0 = \frac{m+1}{2\pi} \cos^m(\phi) \qquad (4.3)$$

where $m = \frac{-\ln 2}{\ln\left(\cos^m(\phi_{1/2})\right)}$ and $\phi_{1/2}$ is the radiation semi-angle. Moreover, in 4.2, $T(\varphi)$ and $g(\varphi)$ are the filter and concentrator channel responses depending on the incidence angle. It is worth noticing that the optical channel depends non-linearly on parameters such as the incidence angle, which affects as a cosine fiction, or the response of the filter and concentrator lens. As discussed in the following section, this feature of the optical channel is extremely useful for reducing the correlation among channel responses, which allows us to maximize the performance of the multiple-input multiple-output (MIMO) schemes when applied to VLC.

Modelling the diffuse component typically requires to define a ray-tracing model and the integration along the surface where the reflection occurs to determine its value. In this sense, the most common model corresponds to the one-bounce scheme defined in [12], in which the frequency channel response

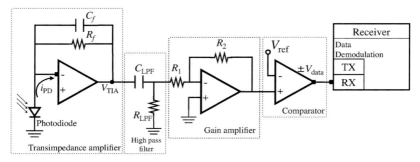

Figure 4.5 General architecture of a VLC receiver.

of the diffuse component is given by [13]

$$h_{\text{diff}} = \frac{\eta_{\text{diff}}}{1 + j\frac{f}{f_d}} \tag{4.4}$$

where f_d is the cut-off frequency of the diffuse contribution. Moreover, the parameter η_{diff} is given by

$$\eta_{\text{diff}} = \frac{A}{A_{\text{room}}} \cdot \frac{\rho_1}{1 - \rho}, \tag{4.5}$$

where A_{room} is the area of the room in which the signal suffers the reflection, ρ_1 is the reflectivity of the surface and ρ is the average reflectivity of the walls.

At this point, it is necessary to consider limited modulation bandwidth of the LEDs. The channel response of a generic LED is denoted by $\eta_{\text{LED}}(t)$ and its frequency response can be simply modeled as $h_{\text{LED}}(f) = e^{-\frac{f}{1.44f_m}}$, where f_m is the -3 dB modulation bandwidth of the LED. Taking into consideration all these contribution, the resulting optical channel can be written as

$$\eta_{\text{eq}}(t) = (\eta_{\text{LOS}} + \eta_{\text{diff}}(t - \Delta T)) * \eta_{\text{LED}}(t), \tag{4.6}$$

Then, the channel response in the frequency domain is given by

$$h_t(f) = \left(\eta_{\text{LOS},t} + \eta_{\text{diff}} \frac{e^{j2\pi\Delta T}}{1 + j\frac{f}{f_d}} \right) e^{-\frac{f}{1.44f_m}}. \tag{4.7}$$

4.2.3 Receiver

The receiver architecture described above corresponds to a simple implementation. It is worth remarking that, based on a similar methodology, more complex electrical models can be considered depending on the characteristics of the VLC system. However, it is worth noticing that, following the proposed design, each photodiode is equipped with a signal processing chain, i.e., a set of electrical components that converts the received optical power into a signal that can be the input of the analog-to-digital converter (ADC) of the device that demodulates and decodes the data. However, as it will be discussed in the following section, assuming that each photodiode is equipped with a specific signal processing chain may lead to an excessive power consumption for multiple photodiode configurations.

4.3 User-centric Approach and Hybrid VLC/RF Networks

Optical APs are characterized by a small and confined coverage footprint comprising few meters, typically referred to as attocell. Moreover, multiple LEDS are typically deployed in the ceiling for providing satisfactory illumination. In this sense, the axiomatic concept of cell that have paved the cellular communications based on managing the intercell interference through frequency reuse results futile in VLC due to the small coverage provided by each optical AP. The user-centric approach for VLC was initially proposed in [27]. Basically, the set of optical APs are grouped forming an optical cell composed of multiple APs each depending on the distribution of the users and their channel responses as it is shown in Figure 4.6. Furthermore, this approach results also useful for adapting the network to the blocking and shadowing effects at the receiver side.

Once the optical cells have been formed, notice that it is necessary to manage the interference between users in the same cells, referred to as intracell interference, and also the intercell interference [17]. Since each cell is composed of several LED transmitters serving multiple users, MIMO transmission schemes can be potentially implemented to manage the intercell interference. On the other hand, although the clustering methods inherently minimize the intercell interference, resource management or alternative interference cancellation schemes can be required [23].

From a networking perspective, VLC are typically deployed in scenarios where a RF systems is already available, e.g., a WiFi AP or 5G and 6G cellular coverage. Furthermore, the integration of VLC in the cellular network, usually referred to as LiFi [26], allows us to exploit the network redundancy. That is, a user that suffers a degradation of the VLC link may handover to the RF network. Similarly, a user connected to the RF network can move to the VLC network when suffering low data rates. In this sense, RF systems

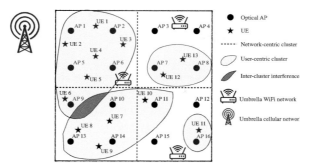

Figure 4.6 Optical cells composed of multiple optical APs following a user-centric approach. Moreover, cellular and WiFi networks allows us to exploit a hybrid RF/VLC network.

can be managed as an umbrella network within a VLC user-centric scenario. It is worth noticing that the performance of the RF networks depends on the implemented technology, e.g., the latest versions of the 802.11 standard for WiFi or 5G and 6G for cellular networks. In fact, the standardization of LiFi in the 802.11bb standard already considers the interaction with other WiFi frequencies, e.g., similarly as it is possible to switch between WiFi at 2.4 GHz and 5 GHz.

At this point, implementing the user-centric approach requires to apply a clustering algorithm and managing the transmission resources of the networks. Typically, clustering methods such as the K-means algorithm have been proposes for obtaining optical cells as a function of the users distribution. However, this clustering methodology simply considers the network topology. Therefore, more complex clustering and optimization methodologies that consider aspects such as the probability of blocking, the diffuse component due to reflections of the optical signal or the load balance between cells must be derived for exploiting the potential of the user-centric approach. As a consequence, a large amount of parameters beyond the Lambertian model are involved for solving issues such as blocking and shadowing or managing the transmission resources. In fact, it is extremely complex to obtain useful solutions through closed-form expressions or traditional complex problems. Recently, in line with the development of 6G communications, artificial intelligence (AI) comprising machine, deep, reinforced, federated, etc. learning algorithms has been proposed for VLC network [6, 25]. Therefore, the application of AI to VLC networks is currently an issue to solve taking into consideration the difference between RF and optical transmission.

4.4 Visible Light Communications for 6G

4.4.1 The concept of MIMO applied to VLC

The 5G and 6G wireless communications exploit the concept of massive MIMO. Basically, increasing the number of transmitters (antennas) regarding the number of users lead to achieving multiplexing gain, i.e., transmitting multiple data streams simultaneously over the same transmission resource avoiding the interference among them, and array gain. At this point, let us define the channel matrix as

$$\mathbf{H} = \begin{bmatrix} h_{1,1} & h_{1,2} & \cdots & h_{1,M} \\ \vdots & \vdots & \ddots & \vdots \\ h_{N,1} & h_{N,2} & \cdots & h_{N,M} \end{bmatrix} \in \mathbb{C}^{N \times M}, \tag{4.8}$$

where $h_{n,m}$ is the channel response between transmitter m, $m = \{1, \cdots M\}$, and receiver n, $n = \{1, \cdots N\}$. Thus, assuming independent and identically distributed (i.i.d.) inputs of the matrix, the following condition hold,

$$\lim_{M \to \infty} \det\left(\mathbf{H}\mathbf{H}^H\right) = M\mathbf{I}_N. \tag{4.9}$$

That is, the multiplexing gain is proportional to the number of antennas. However, to achieve this gain, the channel matrix (4.8) must be well-conditioned, i.e., the channel responses between users are uncorrelated [10, 3]. At this point, the following question arises; *Can the principles of massive MIMO be applied to VLC?* That is, achieving multiplexing and array gain by increasing the number of optical APs in the same area as it is shown in Figure 4.7. It is worth recalling that the concept of massive MIMO does not involve increasing the transmitted power, i.e., the power consumption as the number of transmitters (optical APs) get larger. For instance, assuming a transmission power equal to P_{opt} for a single optical AP and uniform power allocation, a massive MIMO deployment would consider M optical APs with a transmission power of $\frac{P_{\text{opt}}}{M}$ each.

In RF systems, rich scattering environment guarantees well-conditioned channel matrices. That is, the small-scale effects such as Rayleigh fading lead to non-linear channel responses. However, the absence of small scale effects and the fact that the LoS contribution mainly depends on the geometry between transmitter and photodiode (see (4.2)) hampers the achievement of linearly independent channel responses. Besides, note that the well-known Shannon's capacity equation cannot be applied to IM/DD schemes such as VLC. In fact, there not exists any closed-form of the capacity for VLC. In

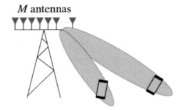

(a) Traditional cellular base station   (b) Massive MIMO approach.

Figure 4.7   Traditional base station concept vs. massive MIMO in RF systems.

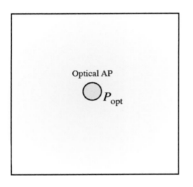

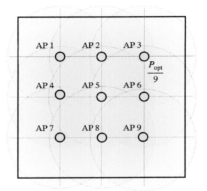

(a) Traditional cellular base station   (b) Massive MIMO approach.

Figure 4.8   Traditional and topological approaches for assigning each user to a specific tier.

order to solve this issue, outer, lower and upper bounds of the capacity have been derived instead [15, 24].

The signal transmitted by the M optical APs is given by the vector $\mathbf{u} = [u_1, \ldots, u_N]^T \in \mathbb{R}^{N \times 1}$ where u_n is the signal corresponding to the l-th transmitter. Thus, the signal received by user n is

$$y^{[n]} = \left(\mathbf{h}^{[n]}\right)^T \mathbf{u} + z^{[n}, \tag{4.10}$$

where $\mathbf{h}^{[n]} \in \mathbb{R}^{1 \times M}$ is the row vector that contains the channel response between the M optical APs and user n and $z^{[n]}$ is real additive white Gaussian noise with variance (AWGN) σ_z^2. Moreover, all the optical APs converge to a central unit (CU) enabling cooperation among them.

Implementing the concept of (massive) MIMO in VLC is based on implementing a precoding scheme. That is, the channel state information (CSI) must be known at the transmitter side for calculating the precoding matrices that cancel or minimize the interference between users served in the same transmission resource, time and frequency. In RF systems, time division duplex (TDD) is assumed for estimating the CSI through the uplink while exploiting the reciprocity between uplink and downlink to obtain the CSI at the transmitters. However, VLC are inherently operating in frequency division duplex (FDD) since uplink is typically implemented in the infrared domain or through a RF system [5].

Taking into consideration the introduction of a precoding of the transmitted signal, let us define the symbol intended to user n as s_n. The vector $\mathbf{s} = \left[s^{[1]}, \ldots, s^{[N]}\right]^T \in \mathbb{C}^{N \times 1}$ contains the symbols to the N users and each symbol s_n is zero-mean corresponding to a value within $[-1, 1]$. Denoting the precoding vector associated to user n as $\mathbf{w}^{[n]} \in \mathbb{R}^{L \times 1}$, the received signal for user m is given by

$$y^{[n]} = \underbrace{\mathbf{h}^{[n]} \mathbf{n}^{[n]} s^{[n]}}_{\text{desired symbol}} + \underbrace{\mathbf{h}^{[n]} \mathbf{I}_{\text{DC}}}_{\text{illumination}} + \underbrace{\mathbf{h}^{[n]} \sum_{i=1, i \neq n}^{N} \mathbf{w}^{[i]} s^{[i]}}_{\text{interference}} + \underbrace{z^{[n]}}_{\text{noise}}, \quad (4.11)$$

where $\mathbf{I}_{\text{DC}}$ is the DC-bias current at each LED transmitter that satisfies the illumination requirements. For the proposed system model, the global channel matrix can be written as $\mathbf{H} = \left[\mathbf{h}^{[1]T} \quad \ldots \quad \mathbf{h}^{[N]T}\right]^T \in \mathbb{R}^{N \times M}$ and the precoding matrix as $\mathbf{W} = \left[\mathbf{w}^{[1]}, \ldots, \mathbf{w}^{[N]}\right] \in \mathbb{R}^{M \times N}$, this condition implies $\mathbf{HW} = \text{diag}\left(\sqrt{\lambda_k}\right)$.

Although the closed-form expression of the capacity does not exist, in [24] a lower bound is derived for multiuser MIMO systems considering transmit precoding schemes. Specifically, omitting the azimuthal and elevation angles, the lower bound of the capacity for user n is

$$C^{[n]} \geq \frac{1}{2} \log \left(1 + \frac{2|\mathbf{h}^{[n]} \mathbf{w}^{[n]}|^2}{\pi e \left(\frac{1}{3} \sum_{i \neq k} |\mathbf{h}^{[n]} \mathbf{w}^{[i]}|^2 + \sigma_z^2\right)}\right). \quad (4.12)$$

At this point, the precoding vectors that achieve a data-rate close to the capacity must be calculated. For illustrative purposes, let us consider a precoding scheme that completely cancels the multi-user interference, usually referred to as zero-forcing (ZF). Again, the closed form expressions typically applied to RF systems do not hold for VLC because of their particular characteristics given by IM/DD, i.e., real and non-negative transmitted signal [22]. Thus, determining the precoding vectors as the pseudoinverse of the channel matrix, i.e., $\mathbf{W} = \mathbf{H}^H \mathbf{H} \mathbf{H}^H$, is not optimal. For VLC, calculating the precoding vector for MIMO-VLC are given by solving the an optmization problem such as

$$\underset{\mathbf{w}^{[n]}}{\text{maximize}} \quad \sum_{n=1}^{N} \frac{1}{2} \log \left(1 + \frac{2|\mathbf{h}^{[n]} \mathbf{w}^{[n]}|^2}{\pi e \sigma_z^2} \right)$$

$$\text{subject to} \quad \mathbf{H}\mathbf{W} = \text{diag} \left(\sqrt{\lambda_n} \right) \tag{4.13}$$

$$\sum_{n=1}^{N} |\mathbf{e}_l \mathbf{w}^{[n]}| < \Delta I_{\text{tx}}, \quad m = 1, \dots, M$$

where $\mathbf{e}_l$ is the unit row vector whose m-th entry is 1 and $\Delta I_{\text{tx}} = \min(I_{\text{DC}}, I_{\text{max}} - I_{\text{DC}})$ is the input current range defined by the maximum current of the optical transmitter, which is denoted by I_{max}. Notice that other precoding strategies such as minimizing the mean square error (MMSE) can be formulated straightforwardly following the same optimization problem structure [16].

Beyond the need for calculating the precoding vectors, the achievable data-rate of (massive) MIMO in VLC mainly depends on the well or ill condition of the channel matrix. Although VLC are not subject to small-scale effects that generate a rich scattering environment, which ensures a well-conditioned channel matrix, the optical channel response allows us to obtain non-linear channel responses by managing the incidence angle or the response of the filter plus concentrator lens. Thus, considering the orientation angle of each user, which is defined by the azimuthal and polar angles denoted $\alpha^{[n]}$ and $\theta^{[n]}$ for user n, respectively, the channel matrix can be rewritten as

$$\mathbf{H} = \left[\mathbf{h}^{[1]} \left(\alpha^{[1]}, \theta^{[1]} \right)^T \quad \cdots \quad \mathbf{h}^{[K]} \left(\alpha^{[N]}, \theta^{[N]} \right)^T \right]^T \in \mathbb{R}^{K \times L} \tag{4.14}$$

At this point, we describe multiple alternative receiver architecture for exploiting the characteristics of the VLC channel to enhance the performance of MIMO schemes.

4.4.2 Angle diversity receivers

Angle diversity receivers or (ADR) is one of the methods that enhance the performance in VLC system and mitigate the interferences that comes from other light sources. For instance, in a single photodiode receiver, using a wide FoV will not only increase the area of dectection of receiver but also increase the interference. Also, the channel gain increases with narrow FoV and vise vera as it is clear from the gain of the optical concentrator. Thus, there exists a trade-off problem between resulting FoV and receiving gain. Fortunately, applying the concept of ADR it is possible to achieve wide FoV and large optical at the same time [11]. In ADR, instead of using a single photodiode, narrow FoV detectors are combined in which each photodiode points to a distinct direction then based on an orientation vector, a photodiode (or a combination of them) is selected according to a predefined criterion. The PDs are analyzed as a MIMO channel system.

ADR is proposed in some research e.g., pyramid photodiode arrangement (PRs) and truncated pyramid arrangement (TPRs) are proposed to enhance the performance of indoor MIMO-VLC system and reduce the signal to inter-ference plus noise ratio (SINR) fluctuation respectively [12] [13]. According to [14], the accuracy of the performance in the ADR system depends on three factors. First, the orientation of the user, the direct current (DC) channel gain significantly affected by the orientation of the user, and random orientation should be considered in the ADR system instead of assuming the receiving device is pointed vertically upward. Second, diffuse link signal propagation, in indoor scenario some signals are reflected from walls or any other object, so non-line of sight (NLOS) should not be neglected in ADR. Third, noise power spectral density where system performance is significantly affected by the noise power spectral density of the photodiode.

To understand the concept of the ADR, we assume the pyramid receivers (PRs) that shown in Figure 4.2, azimuthal and elevation angles of the user k of the photodiode ν are α, δ respectively. Orientation vector $\hat{\mathbf{n}}_k$ is given by,

$$\hat{\mathbf{n}}_k = \left[\sin(\theta_k(\nu))\cos(\alpha_k(\nu)), \quad \sin(\theta_k(\nu))\sin(\alpha_k(\nu)), \quad \cos(\theta_k(\nu))\right].$$
(4.15)

Then,
the irradiation and the incident angles can be calculated using the following equations:

$$\phi_{kt} = \arccos\left(\frac{(\hat{\mathbf{n}}_t \cdot \mathbf{v}_{tk})}{\|\hat{\mathbf{n}}_t\|\|\mathbf{v}_{tk}\|)}\right),$$
(4.16)

and

$$\varphi_{kt}(\nu) = \arccos\left(\frac{(\mathbf{v}_{tk} \cdot \hat{\mathbf{n}}_k(\nu))}{\|\mathbf{v}_{tk}\|)\|\hat{\mathbf{n}}_k(\nu)\|}\right)$$
(4.17)

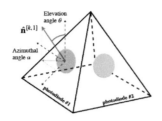

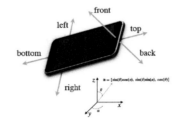

(a) ADR in a 4-face pyramidal arrangement.

(b) Omnidirectional FoV
ADR in hald-held device.

Figure 4.9 ADR architectures and their implementation in hand-held devices.

where, ϕ_{kt} and $\varphi_{kt}\nu$ are the irradiation and the incident angles respectively. Moreover, $\hat{n}_t$, v_{kt} are the pointing vector of the LED t and the vector generated from the LED t to user k.

As described in [ref], there are multiple geometrical configurations to implement the ADR For pyramidal arrangements, the photodiodes are uniformly distributed in a circle of radius r on the x-y horizontal plane as it is shown in Figure 4.3. Assuming N photodiodes at each receiver and $\nu = \{1, \ldots, N\}$, the coordinates on photodiode nu is given by,

$$(x_k(\nu), y_k(\nu), z_k(\nu)) = \left(x_{pr} + r \cdot \cos\left(\frac{2(\nu-1)\pi}{N}\right), \right.$$
$$\left. y_{pr} + r \cdot \sin\left(\frac{2(\nu-1)\pi}{N}\right), h_{pr} \right) \quad (4.18)$$

where, x_{pr}, y_{pr} are the x-y coordinates measured from the center of the pyramidal. h_{pr} is the height of the photodiode. The orientation of the photodiode is identified by the azimuthal and elevation angles denoted as $\alpha_k(\nu) = \left(\frac{2(\nu-1)\pi}{N}\right)$ and $\theta_k(\nu) = \theta_k$ respectively. For the hemispherical arrangement, the polar and azimuthal angles are given by

$$\theta_i = \arccos(t_i) \quad (4.19)$$

for $1 \leq i \leq N$ and for $2 \leq i \leq N$,

$$\phi_i = \left(\phi_{i-1} + \frac{3.6}{\sqrt{N}} \frac{1}{\sqrt{1 - t_i^2}} \right) \pmod{2\pi} \quad (4.20)$$

where $\phi_1 = 0$ and $t_i = 1 - \frac{2(i-1)}{2N-1}$.

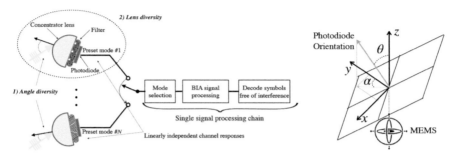

(a) Architecture of the reconfigurable photodetector   (b) Architecture of the vROA

Figure 4.10   Architectur of the reconfigurable and vROA photodetectors.

Furthermore, the receivers are also subject to to movement in the three Euler angles; pitch, roll and yaw. Focusing on hand-held devices, omnidirectional FoV can be achieved simply allocating a photodiode in each face of the device as it is shown in Figure 4.9(b).

4.4.3 Reconfigurable photodetector

The ADR architectures typically assume a signal processing chain as described in Figure 4.5 per photodiode. In this sense, the complexity and power consumption of the previous ADRs grows linearly with the number of photodiodes. In [20], the architecture of reconfigurable photodetector is proposed, in which a set of photodiodes in angular and also lens diversity is connected to a single signal processing chain through a selector. Thus, the reconfigurable photodetector selects a specific channel response at each time. This architecture approach results similar to the analog/digital architecture proposed for RF systems in order to implement hybrid precoding. In fact, it obtains the same benefits, selecting proper signal responses while reducing the number of amplifier and ADCs, which have a direct impact reducing the energy consumption.

First notice that the same geometrical arrangements, e.g., pyramidal, hemispherical, etc., as for ADRs can be applied to the reconfigurable photodetector architecture, which leads to non-linear responses since the channel varies as the cosine function of the incidence angle. Moreover, filter and concentrator lens diversity is assumed to obtain linearly independent responses between the modes (photodiodes) of the reconfigurable photodetector of each user. Specifically, considering a lens architecture as described in Figure 4.11, the channel response because of the lens/concentrator of each photodiodes is

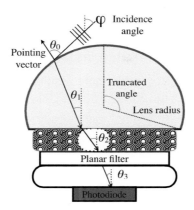

Figure 4.11 Propagation of the optical signal through a generic lens.

given by,

$$g(\varphi) = \frac{\int_{S_0} T(\theta_0)T(\theta_1)T(\theta_2)T(\theta_3)dS}{\int_{S_0} \cos(\theta_0)dS}, \qquad (4.21)$$

where S_0 is the surface for which light passing finally hitting the photodiode and $T(\theta)$ is the optical gain of the lens. This propagation model is depicted in Figure 4.11. Notice that the lens design may provide multiple possibilities for generating linearly independent channel responses. The particular case of an hemispherical concentrator ($\theta_t = 90°$) provides a channel response equal to

$$g(\varphi) = \begin{cases} \frac{n_s^2}{\sin^2 \Psi_c} & \text{if } 0 \le \varphi \le \Psi_c \\ 0 & \text{if } \varphi \le \Psi_c. \end{cases} \qquad (4.22)$$

where n_s is the intern reflective index achieves of the lens.

The reconfigurable photodetector was initially proposed for implementing blind interference alignment schemes (BIA), a signal processing technique that achieves multiplexing gain without the need for channel state information (CSI) at the transmitter or cooperation among them in MIMO-VLC [20]. Beyond this applications, the reconfigurable photodetectors have been also proposed for improving the cell formation in user-centric approaches. That is, this receiver architectures is also useful for managing the intercell interference and managing the transmission resource. For MIMO precoding it is worth noticing that the reconfigurable photodetectors allow us to select proper channel responses for each user. That is, optimizing the channel matrix to maximize the orthogonality among channel responses of the users.

4.4.4 Variable receiving orientation angle devices

The reconfigurable photodetector architecture requires to deploy a set of photodiodes in each device. With the aim of avoiding this exceed of photodiodes, the variable receiver orientation angle (vROA) architecture considers a single photodiode whose orientation can be varied through microelectromechanical systems (MEMS) [19] as it is shown in Figure 4.10(b). Similarly to the photodiode selection for the reconfigurable photodetector, the vROA can select N_ν orientations determined by the pointing accuracy of the MEMS. Denoting the polar and azimuthal accuracy as $\Delta\theta$ and $\Delta\alpha$, respectively, the vROA architecture can select a channel response among $N_\nu = \frac{180°}{\Delta\theta} \cdot \frac{360°}{\Delta\alpha}$ orientation, e.g, 64800 orientations are available assuming an accuracy of one degree.

Focusing on the application of MIMO precoding in VLC, the vROA architecture allows us to maximize the orthogonality of space generated by the channel responses of the users. In [19], it is demonstrated that a pointing accuracy of $\Delta\theta = \Delta\alpha = 10°$ achieves a ZF sum-rate close to the capacity. Interestingly, the vROA architecture has been aslso proposed for selecting the channel matrix instead of estimating it. That it, assuming that the transmitters know the topology of the network, the optimal orientation are calculated and, after that, they are specified to the users through the downlink.

4.4.5 Intelligent reconfigurable surfaces

Recently, intelligent reconfigurable surfaces (IRS) have been considered for increasing the degrees of freedom of the propagation environment for RF systems [18]. Interestingly, VLC may benefit of the concept of IRS since LoS propagation and blocking or shadowing effects play a major role in VLC. In this sense, there are two IRS technologies available in VLC; metasurfaces and mirror arrays [2, 1].

Metasurfaces are based on exploiting the light propagation and reflection of synthesized materials composed of arrangements of sub-wavelength metallic or dielectric structures. These material are similar to the surfaces that compose the optical lenses. Thus, the metasurfaces can modify the wavelength, polarization and phase of the received signal as well as split or unify optical beams. On the other hand, mirror arrays are composed of, N_p^2,

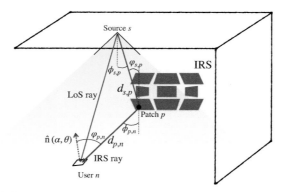

Figure 4.12 Propagation of the optical signal in a IRS environment.

$n_p \times n_p$, patches, also referred to as micromirrors, that reflect the optical signal following the general Snell's law as it is shown in Figure 4.12. At this point, the luminous power generated by a nearby LED is received with a uniform radiation emittance over the area of the IRS. Similarly to the vROA photodetector, each micromirror selects a specific orientation angle employing a MEMS. Assuming the Lambertian model, each patch generates a new signal ray with a specific radiation angle ϕ_p, which can be obtained applying the Snell's law, and it is received with an incidence angle φ_p. Therefore, applying the Lambertian model the channel received from the p-th patch is given by

$$\eta_p = h_{\text{source}\to\text{patch}} \cdot h_{\text{patchs}\to\text{user}} \\ = \delta \cdot \eta_{\text{LoS}}\left(d_{s,p}, \phi_{s,p}, \varphi_{s,p}\right) \cdot \eta_{\text{LoS}}\left(d_{p,n}, \phi_{p,n}, \varphi_{p,n}\right), \tag{4.23}$$

where δ is the reflectin efficiency of the micromirrors and $d_{s,p}$, $\phi_{s,p}$ and $\varphi_{s,p}$ are the distance, radiation and incidence angles between the ligth sources and the micromirror p, respectively, and $d_{p,n}$, $\phi_{p,n}$ and $\varphi_{p,n}$ are the distance, radiation and incidence angles between the micromirror p and user n. Recall that the Lambertian channel is given by (4.1). Therefore, the resulting channel for the receiving user is given by

$$h = \eta_{\text{LoS}}\left(d, \phi, \varphi\right) + \sum_{p=1}^{N_p^2} \eta_p. \tag{4.24}$$

4.4.6 Impact of angle and lens diversity in MIMO-VLC

So far, alternative receiver architectures have been described, which basically are based on increasing the degrees of freedom of the propagation environment, i.e., the number of paths for the transmitted signal. Thus, each user can select the channel responses that minimize the correlation among users. At this point, it seems that the concept of massive MIMO can be applied in VLC exploiting these architectures.

The achievable sum-rate for different receiver architectures is depicted in Figure 4.13. Recall that the sum of the optical power of all the transmistters is the same for comparison purposes. Assuming users equipped with a traditional photodetector composed of a single photodiode, it can be seen that greater sum-rate is achievable as the number of transmitters increases within a reduced range of number of transmitters while it slightly increases for a high number of transmitters. That is, the probability of obtaining correlated channel responses is much higher for a large number of transmitters in VLC. Interestingly, similar behaviour is obtained when implement multiple photodiode configurations, i.e., the achievable sum-rate increases for a reduced number of transmitters while slightly increases for deployments comprising a large number of transmitters. However, there exists an offset between single and multiple photodiodes configurations. In fact, it is shown that the use of alternative receiver architectures achieves a sum-rate close to the channel capacity for VLC. Analyzing the capacity outer bound (see (4.12)), notice

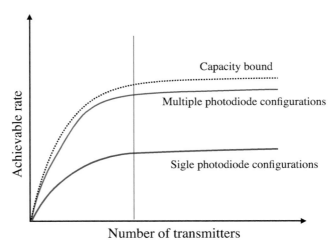

Figure 4.13   Visible light spectrum.

that the capacity slightly increases for large number of transmitters indeed. That is, beyond some density of transmitters, it is not possible to differentiate the signal optical received from neighbouring optical APs. In other words, the capacity of MIMO schemes does not scale for VLC.

4.5 Conclusions

In this chapter, we have focused on the motivation of implementing VLC in future wireless communications considering alternative receiver configurations and materials. First, the elements that composed a VLC systems have been described paralleling with the components of RF systems such as beam management, number of signal processing chains per photodiode/antenna or the particular modulation constraints of VLC. After that, alternative receiver architectures are presented named angle diversity receiver, reconfigurable photodetector, variable receiving orientation angle and intelligent reconfigurable surfaces. it is shown that theses configurations allow us maximizing the performance of the existing MIMO precoding schemes. However, it shown that taking into consideration the optical propagation, even with the proposed alternative configurations, the MIMO capacity does not scales as the number of transmitters grows from some specific density of optical APs in the area of interest.

References

[1] Amr M. Abdelhady, Ahmed K. Sultan Salem, Osama Amin, Basem Shihada, and Mohamed-Slim Alouini. Visible light communications via intelligent reflecting surfaces: Metasurfaces vs mirror arrays. *IEEE Open Journal of the Communications Society*, 2:1–20, 2021.

[2] Hanaa Abumarshoud, Lina Mohjazi, Octavia A. Dobre, Marco Di Renzo, Muhammad Ali Imran, and Harald Haas. Lifi through reconfigurable intelligent surfaces: A new frontier for 6G? *IEEE Vehicular Technology Magazine*, 17(1):37–46, 2022.

[3] Emil Björnson, Jakob Hoydis, and Luca Sanguinetti. Massive MIMO has unlimited capacity. *IEEE Transactions on Wireless Communications*, 17(1):574–590, 2018.

[4] Nan Chi, Yingjun Zhou, Yiran Wei, and Fangchen Hu. Visible light communication in 6g: Advances, challenges, and prospects. *IEEE Vehicular Technology Magazine*, 15(4):93–102, 2020.

[5] Junil Choi, David J. Love, and Patrick Bidigare. Downlink training techniques for FDD massive MIMO systems: Open-loop and closed-loop training with memory. *IEEE Journal of Selected Topics in Signal Processing*, 8(5):802–814, 2014.

[6] Bekir Sait Ciftler, Abdulmalik Alwarafy, and Mohamed Abdallah. Distributed DRL-based downlink power allocation for hybrid RF/VLC networks. *IEEE Photonics Journal*, 14(3):1–10, 2022.

[7] European Commission, Joint Research Centre, G Zissis, P Bertoldi, and T Serrenho. *Update on the status of LED-lighting world market since 2018*. Publications Office, 2021.

[8] Máximo Morales Céspedes, Borja Genovés Guzmán, Víctor P. Gil JimÃſnez, and Ana García Armada. Aligning the light for vehicular visible light communications: High data rate and low-latency vehicular visible light communications implementing blind interference alignment. *IEEE Vehicular Technology Magazine*, 18(1):59–69, 2023.

[9] Svilen Dimitrov and Harald Haas. *Principles of LED Light Communications: Towards Networked Li-Fi*. Cambridge University Press, 2015.

[10] Jakob Hoydis, Stephan ten Brink, and Merouane Debbah. Massive MIMO in the UL/DL of cellular networks: How many antennas do we need? *IEEE Journal on Selected Areas in Communications*, 31(2):160–171, 2013.

[11] Yuhong Huang, Jing Jin, Mengting Lou, Jing Dong, Dan Wu, Liang Xia, Sen Wang, and Xiaozhou Zhang. 6G mobile network requirements and technical feasibility study. *China Communications*, 19(6):123–136, 2022.

[12] V. Jungnickel, V. Pohl, S. Nonnig, and C. von Helmolt. A physical model of the wireless infrared communication channel. *IEEE Journal on Selected Areas in Communications*, 20(3):631–640, 2002.

[13] J.M. Kahn and J.R. Barry. Wireless infrared communications. *Proceedings of the IEEE*, 85(2):265–298, 1997.

[14] Dilukshan Karunatilaka, Fahad Zafar, Vineetha Kalavally, and Rajendran Parthiban. LED based indoor visible light communications: State of the art. *IEEE Communications Surveys & Tutorials*, 17(3):1649–1678, 2015.

[15] Amos Lapidoth, Stefan M. Moser, and MichÈle A. Wigger. On the capacity of free-space optical intensity channels. *IEEE Transactions on Information Theory*, 55(10):4449–4461, 2009.

[16] Baolong Li, Jiaheng Wang, Rong Zhang, Hong Shen, Chunming Zhao, and Lajos Hanzo. Multiuser MISO transceiver design for indoor downlink visible light communication under per-LED optical power constraints. *IEEE Photonics Journal*, 7(4):1–15, 2015.

[17] Xuan Li, Rong Zhang, and Lajos Hanzo. Optimization of visible-light optical wireless systems: Network-centric versus user-centric designs. *IEEE Communications Surveys & Tutorials*, 20(3):1878–1904, 2018.

[18] Yuanwei Liu, Xiao Liu, Xidong Mu, Tianwei Hou, Jiaqi Xu, Marco Di Renzo, and Naofal Al-Dhahir. Reconfigurable intelligent surfaces: Principles and opportunities. *IEEE Communications Surveys & Tutorials*, 23(3):1546–1577, 2021.

[19] Máximo Morales-Céspedes, Harald Haas, and Ana García Armada. Optimization of the receiving orientation angle for zero-forcing precoding in VLC. *IEEE Communications Letters*, 25(3):921–925, 2021.

[20] Máximo Morales-Céspedes, Martha Cecilia Paredes-Paredes, Ana García Armada, and Luc Vandendorpe. Aligning the light without channel state information for visible light communications. *IEEE Journal on Selected Areas in Communications*, 36(1):91–105, 2018.

[21] Parth H. Pathak, Xiaotao Feng, Pengfei Hu, and Prasant Mohapatra. Visible light communication, networking, and sensing: A survey, potential and challenges. *IEEE Communications Surveys & Tutorials*, 17(4):2047–2077, 2015.

[22] Thanh V. Pham and Anh T. Pham. Coordination/cooperation strategies and optimal zero-forcing precoding design for multi-user multi-cell vlc networks. *IEEE Transactions on Communications*, 67(6):4240–4251, 2019.

[23] Ahmad Adnan Qidan, Máximo Morales-Céspedes, Ana García Armada, and Jaafar M. H. Elmirghani. Resource allocation in user-centric optical wireless cellular networks based on blind interference alignment. *Journal of Lightwave Technology*, 39(21):6695–6711, 2021.

[24] Hong Shen, Yuqin Deng, Wei Xu, and Chunming Zhao. Rate maximization for downlink multiuser visible light communications. *IEEE Access*, 4:6567–6573, 2016.

[25] Liqiang Wang, Dahai Han, Min Zhang, Danshi Wang, and Zhiguo Zhang. Deep reinforcement learning-based adaptive handover mechanism for VLC in a hybrid 6G network architecture. *IEEE Access*, 9:87241–87250, 2021.

[26] Xiping Wu, Mohammad Dehghani Soltani, Lai Zhou, Majid Safari, and Harald Haas. Hybrid LiFi and WiFi networks: A survey. *IEEE Communications Surveys & Tutorials*, 23(2):1398–1420, 2021.

[27] Rong Zhang, Jiaheng Wang, Zhaocheng Wang, Zhengyuan Xu, Chunming Zhao, and Lajos Hanzo. Visible light communications in heterogeneous networks: Paving the way for user-centric design. *IEEE Wireless Communications*, 22(2):8–16, 2015.

[28] Yang Zhao, Jun Zhao, Wenchao Zhai, Sumei Sun, Dusit Niyato, and Kwok-Yan Lam. A survey of 6G wireless communications: Emerging technologies, 2020.

5

Access Security in 6G: The 6G-ACE Protocol (A Concept Proposal)

Geir M. Køien

University of South-Eastern Norway (USN), Norway

Abstract

Native access security in 5G is based on an enhanced Authentication and Key Agreement (AKA) protocol (5G-AKA) and a new identity presentation scheme that provides subscriber identity privacy. The 5G authentication is more-or-less mutual between the home environment and the user equipment, but the solution has some oddities. The 5G-AKA protocol is constrained by the architectural evolution of the system, and the core of the 5G-AKA protocol is dependent on the third generation UMTS AKA algorithms.

In this chapter, we analyze the history, the current situation and what the access security needs of 6G might be. We then present a concept proposal of what 6G subscriber access security protocol could be. The design of the "6G Authentication and Context Establishment"' protocol (6G-ACE) will focus on principles and concepts, while avoiding detail when possible. The solution should be fast, effective, and reliable, and it should provide security contexts tailored to the task at hand. A 6G solution should break with existing solution, when necessary, but it should also seek to retain proven designs when feasible.

Keywords: Mobile Access Security, Mutual Entity Authentication, Identity Privacy, Security Contexts, 6G-ACE

5.1 Introduction and Background

5.1.1 Visions for 6G

The requirements capture for the International Mobile Telecommunications (IMT) systems is done by the International Telecommunication Union (ITU), which is a United Nations agency. Specifically, it is the ITU-R (radiocommunications aspects) and Working Party (WP) 5D, which is responsible for capturing the high-level requirements for the IMT systems. Previously, they have specified the IMT-2000, IMT-Advanced, and IMT-2020 system. These IMT definitions correspond to 3G, 4G, and 5G respectively.

In May 2021, ITU-R WP-5D started work on the "IMT Vision for 2030 and beyond" concept. The recommendation for "IMT for 2030 and Beyond" is expected to be similar in scope to the 5G vision [1]. The 5G recommendation gave us a high-level specification, usage scenarios, etc. The "triangle vision" is from the 5G recommendation.

The work on IMT-2030 is still in draft status, but a few aspects are known. Among the known aspects are ITU-R Report M.2516-0 "Future technology trends of terrestrial International Mobile Telecommunications systems towards 2030 and beyond" [2]. This report will be an important input document to the IMT-2030 specification process.

The 3GPP is the organization where the actual standards are being developed. The IMT recommendations serve as input and requirements to the 3GPP process. Currently, the 3GPP is about to define *Release 18* of the system. Release 18 is also known as "5G Advanced" and as "5.5G." Release 18 is expected to be finalized in mid-2024. Many of the features and functionalities of Release 18 will be carried over to the initial phases of 6G. We shall return to the technical expectations for 6G in clause 5.4.

5.1.2 Access security and identity confidentiality

Access security in the mobile systems has evolved and improved over the years. Yet, some elements are still dependent on design decisions made for the 1G/2G systems. A case in point is the identity presentation procedure, which needs to be in plaintext (unprotected). This has prohibited credible subscriber identity privacy. The 5G systems has basically retained the UMTS-AKA (3G) authentication scheme. That is, the core of the protocol is still required to be backwards compatible with the 3G USIM. This has benefited backwards compatibility and system migration aspects, but it has made the 5G-AKA protocol more complex.

The 5G system has a "Subscription Concealed Identifier" (SUCI) scheme. The SUCI-scheme is based on use of Elliptic Curve Integrated Encryption Scheme (ECIES). However, the SUCI-scheme is not integrated with the 5G-AKA protocol and technically it is an optional procedure. It is noted that the SUCI scheme (ECIES) is not quantum-safe. Quantum-safe cryptography (QSC) refers to algorithms that are resistant to attacks by both classical and quantum computers. One also uses the term post-quantum cryptography (PQC) to refer to the same concepts.

5.1.3 Goals for a 6G authentication and context establishment protocol

There are many possible high-level goals for a 6G-ACE protocol. We adopt the stance that the protocol should be as simple as possible. To this end, the protocol should avoid features that are not strictly required. It should also avoid legacy dependencies that aren't future proof. Thus, the 6G-ACE protocol should avoid UMTS-AKA/USIM dependencies when possible.

The 6G-ACE protocol may retain functionality from 5G access security that are regarded necessary, useful, and robust. The derivation of anchor keys and setting up security contexts that reflects the trust relationships may therefore be retained.

Backwards compatibility with previous generations is otherwise considered a non-goal. Realistically, a mobile device capable of supporting the 6G radio access can easily also accommodate separate support for 4G/5G security.

5.2 Principal Entities and Trust Relationships

The following is a generic outline of the types of principal entities in the system. We also define trust relationships, which have been invariant over the generations.

5.2.1 Principal entity types

There are three principal entity types in the mobile systems. These have had several different names over the years. We have the following definitions:

- *The home environment (HE)*
  The HE is the "home operator." It manages administrative and operative subscription data. This includes mobility management information, subscription identifiers, and security credentials.

- *The serving network (SN)*
  The SN contains core network and access network functionality to host roaming subscribers, and it facilitates provisioning of connectivity services. It has a roaming profile of the subscribers that it hosts.

- *The user entity (UE)*
  The UE consists of a mobile equipment (ME) and a subscription identity module (SIM). The SIM contains security credentials, etc. We denote our SIM incarnation as the ACE-SIM (hosted on a trusted platform).

Many operators will have both HE and SN functionality. The HE and SN are nevertheless distinct entities. We define roaming to include cases where the HE and SN are managed by the same operator.

We also assume the presence of a Dolev-Yao intruder (DYI) in the system [3]. The DYI is an archetype intruder and is usually modeled as being the communications medium (here: both wired and wireless). It will store all communication for possible later use, and it can inject, reflect, delete, and modify messages at will. The DYI cannot break cryptographic primitives per se, but it can exploit any inappropriate use of cryptographic measures.

5.2.2 Trust relationships

The long-term trust relationships between the principal entity types are based on contractual agreements. There is also a derived trust relationship for the roaming subscriber. These relationships, as depicted in Figure 5.1, have been a stable property of the mobile systems over the generations:

- *The HE-SN roaming agreement (long-term)*
  This trust relationship is based on a legally binding roaming agreement contract. For our purpose, the trust can be assumed to be symmetric.

- *The HE-UE subscription contract (long-term)*
  The HE is the issuer of UE identifiers and security credentials. For public operators, the HE must adhere to national regulatory requirements. The trust is nevertheless asymmetric, with HE as the dominant principal.

- *The SN-UE-derived relationship (transient, spatio-temporally confined)*
  This relationship is based on the HE-SN and HE-UE relationships. It is established by on-demand, based on the existing security contexts (HE-SN and HE-UE). The relationship is temporary by nature, and only exists for the duration of a roaming period. For prolonged roaming periods, the relationship will need to be refreshed.

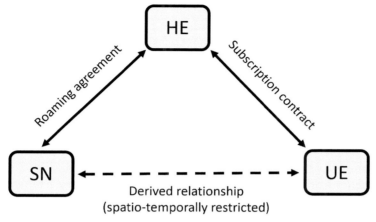

Figure 5.1 The HE, SN, and UE trust relationships.

5.3 History Lessons

The mobile systems have a long history by now. It is worthwhile to take a brief look at the antecedent AKA protocols. We have five existing generations of mobile system, with approximately a decade between them. The AKA protocols have changed during the generations, partially due to different radio access requirements, but also due to requirements to strengthen the security.

5.3.1 The 1G systems

Our example system is the Nordic Mobile Telephony system (NMT), which provided digital call setup and analog speech/audio [4]. Originally, there was no cryptographically based authentication, but NMT eventually got the so-called NMT-SIS scheme. It provided a simple unilateral challenge–response scheme, where the network would have assurance of the subscriber identity. There was no need for key agreement in NMT.

5.3.2 The 2G systems

Our 2G example system is GSM, which is fully digital, and with basic access security capabilities. The GSM system is also the basis for the 3GPP systems. The system is defined in a set of technical specifications (TS).

The GSM system features a unilateral challenge–response protocol (GSM-AKA). The protocol is a two-staged delegated protocol, in which the authentication of the UE is handled by the SN. To carry out this, the SN requests authentication sets (triplets) from the HE. The triplets consist of a random challenge, the associated signed response, and the session cipher key. The GSM-AKA protocol lets the SN to authenticate the UE, but the UE will not have any assurance of the SN or the HE.

$$\text{GSM Triplet} = \{\text{RAND}, \text{SRES}, \text{Kc}\}. \tag{5.1}$$

The GSM system has over-the-air encryption. The cipher interface is called A5, and the key is 64-bit Kc. Over the years there have been several A5 algorithms, the original being the A5/1. The subscriber credentials consist of a permanent identifier, IMSI, and an authentication key (Ki). The IMSI is globally unique and includes information about nationality and the mobile network operator. The 128-bit wide Ki key was introduced with GSM, and it has been retained for all subsequent generation as the primary authentication key (later renamed as K).

$$\text{GSM Credentials} = \{\text{IMSI}, \text{Ki}\}. \tag{5.2}$$

At the UE, the credentials and cryptographic algorithms are stored on the SIM. Originally the SIM was a combined hardware (smartcard) and software module. Currently, the SIM is an application running on the smartcard. The main GSM security specification is TS 43.020 [5]. Both the GSM-AKA and the encryption procedures are optional.

5.3.3 The 3G systems

Our example is the 3GPP-based UMTS system. The UMTS system is all-digital and has reasonable IP support. The UMTS-AKA protocol is modelled on the GSM-AKA protocol, and it retains the two-staged approach. UMTS-AKA also retains the basic challenge–response scheme, but with improvements. The challenge consists of RAND and an authentication token (AUTN). The token includes an authentication tag, a sequence number, and an authentication management field. This provides the UE with assurance that the HE issued the challenge (at some point in time). UMTS also has confidentiality protection (CK, 128-bit key) and integrity protection for the signalling (IK, 128-bit key, 32-bit mac/tag). The authentication set was expanded into an authentication vector (AV).

$$\text{UMTS AV} = \{\text{RAND}, \text{RES}, \text{AUTN}, \text{CK}, \text{IK}\}. \tag{5.3}$$

The SIM from 2G was replaced with a UICC+USIM, where UICC is the physical smartcard and USIM is the "UMTS SIM" application. The IMSI identifier was retained. The Ki was renamed to K but is otherwise the same.

$$\text{UMTS Credentials} = \text{IMSI,K}. \tag{5.4}$$

There are many cryptographic functions, but we shall not elaborate on these here. The interested reader will find more in TS 33.102 [6], which defines the 3G security architecture. An introductory presentation of UMTS access security can be found in [7] and a comprehensive overview over UMTS security can be found in the book "UMTS security" [8].

5.3.4 The 4G system

The 3GPP defined the long-term evolution (LTE) as their 4G system. Technically, the LTE system is considered a 3.9G system, and only properly became 4G with the LTE-Advanced edition. The Evolved Packet System (EPS) is another name for LTE. The system is an All-IP system, and except for backwards compatibility there are no non-IP nodes left in the system.

The LTE system features the EPS-AKA protocol that provides (offline) authenticated challenges. The authentication part of EPS-AKA is almost identical to its UMTS-AKA predecessor. The key agreement part of the protocol is redesigned, and one now derives a key-deriving keys and a full key hierarchy. An EPS-AKA protocol run sets up a security context with a 256-bit master key (K_{ASME}), from which the key hierarchy is derived. The LTE architecture has separated the *control plane (CP)* and *user plane (UP)*, which created a need for more session keys. One has also separate encryption schemes for over-the-air communications and for CP signaling towards the access security management entity (ASME). The ASME is a network function in the SN, and it is the entity that carries out the SN part of the EPS-AKA protocol. The 4G AV reflects the change from providing session keys to providing a key-deriving key.

$$\text{4G AV} = \{\text{RAND}, \text{RES}, \text{AUTN}, K_{\text{ASME}}\}. \tag{5.5}$$

The sessions key are 128-bit wide and key-deriving keys are 256-bit wide. One still uses the IMSI identifier and the corresponding authentication key K. The key K is 128-bit wide, which makes the notion of 256-bit security for the key-deriving keys somewhat contrived.

$$\text{4G Credentials} = \{\text{IMSI}, K\}. \tag{5.6}$$

A peculiarity of the EPS-AKA protocol is that it internally depends on UMTS-AKA functionality. This permits the UE to continue to use a USIM as the basis, and the USIM at the UE still performs the UMTS-AKA protocol. A very good solution for backwards compatibility, but also a solution that limits and restricts the EPS-AKA protocol. The ME therefore still receives RES and (CK,IK) key pair from the USIM. Consequently, it is the ME that must construct the K_{ASME} (from the (CK,IK) keys). This construction includes the SN identifier, which binds the key hierarchy to the specific SN. The USIM is ignorant of the SN-binding. A consequence of the K_{ASME} construction is that the ME has become more important in 4G security than it was with 3G security. The LTE security architecture is specified in TS 33.401 [9]. The book "LTE Security" provides a comprehensive overview [10].

5.3.5 The 5G system

The 5G system architecture deviates substantially from the previous generations. There is a completely redesigned core network architecture and new radio system. The 5G-AKA protocol, while slightly more advanced, is still very similar to its 4G cousin. There are now two anchor keys: a HE anchor key which is called K_{AUSF} and an SN anchor key called K_{SEAF}. The biggest news with 5G access security is the Subscription Concealed Identifier (SUCI) scheme.

The 5G-AKA protocol is in effect an online protocol. That is, while there still is delegation and an AV scheme, the HE needs to be online during the 5G-AKA protocol. Remarkably, one still has the USIM dependency, and the basic UMTS-AKA protocol machinery is embedded within 5G-AKA. Similarly, to the EPS-AKA, the ME needs to carry out all the non-USIM parts of the protocol. The subscriber credentials are more-or-less the same as for LTE, but one now has the Subscription Permanent Identifier (SUPI) as the (generic) subscriber identifier. The SUPI can be instantiated by different types of identifiers, including the IMSI and the network access identifier (NAI) types of identifiers.

For subscriber privacy reasons, the SUPI must not be exposed on the over-the-air interface. To facilitate SUPI privacy, an ephemeral identifier, called SUCI, will be used for the initial UE identity presentation to the SN. Subsequently, the SUCI will be replaced by the 6G version of the Globally Unique Temporary Identifier (GUTI). The HE elliptic-curve Diffie–Hellman public key and parameters are pre-shared with the UE.

$$5\text{G Credentials} = \{\text{SUPI}, \text{K}, \text{ECDH}_{HE}, \text{param}\}. \qquad (5.7)$$

The AV concept is still in use. The K_{ASME} is replaced by the K_{AUSF}, and the response field is somewhat modified. The AUSF is a network function and the security anchor within the HE network. In 5G, there is also a security anchor function in the SN called the SEAF. The key hierarchy has a corresponding key-deriving anchor key for the SEAF is called the K_{SEAF}. It is noted that the primary authentication key K *may* now be 256-bit wide.

$$5\text{G AV} = \{\text{RAND}, \text{XRES}_*, \text{AUTN}, \text{K}_{AUSF}\}. \qquad (5.8)$$

The 5G security architecture is specified in TS 33.501 [11]. The article "3GPP 5G security" provides a nice overview over 5G security [12].

5.3.6 The SUCI scheme

The SUCI scheme is based on asymmetric cryptography, and specifically on use of the Elliptic Curve Integrated Encryption Scheme (ECIES). The SUCI scheme is specified in Annex C in [11]. ECIES is further specified in [13, 14].

The SUCI procedure starts with the UE generating an ephemeral key-pair. The HE public key and the UE private key are used to compute the DH secret, which is subsequently used to derive confidentiality and integrity keys. These are used to protect the "plaintext block" (the SUPI). The SUCI consists of a HE identifier, the protected SUPI, the UE's ephemeral public key, the integrity checksum (called tag) and other parameters. The HE, using the received UE public key, computes the DH secret. It then decrypts the contents and verifies that the SUPI is valid.

5.3.7 Softwareization of the SIM

The 5G system architecture provides a whole new level of virtualization and softwareization. It also includes the eSIM concept, which is a portable software-only realization of the stockticketSIM. The eSIM requires a secure processing platform, called the eUICC. The eUICC may be a smartcard, but can also be integrated directly in the ME. The eSIM may be remotely provisioned and has its own unique eSIM ID (EID). The eSIM is specified by the GSM Association [15].

5.3.8 The evolution

From the 2G systems and onward, there is a well-defined lineage. Newer systems have generally had to cater to the previous generation(s). Design-wise, it has been customary to amend and extend, rather than devise new schemes and procedures. Design and implementation lead times are important considerations, as is backwards compatibility. Deprecation of features happens but tends to take considerable time to execute operationally. When there are multiple parties involved, like when interfacing with subscriber equipment or for inter-system interfaces, backwards compatibility is especially important.

The mobile systems are very clearly evolved systems, even when a clean design would have been preferable. From a security point of view, backwards compatibility tends to be problematic, and it may compromise the security improvements. Backwards compatibility is also a source of added complexity. Complexity will generally complicate designs and implementations and will lead to an increased attack surface.

5.3.9 Mutual entity authentication and perfect forward secrecy

The 5G-AKA protocol provides a somewhat contrived and limited support for mutual authentication. With a clean design, this could easily have been fixed.

Another aspect that has been missing is perfect forward secrecy (PFS). It is not a make-or-break feature, but it would have provided a new level of robustness to the AKA protocols. The PFS property is not a new idea, originating from a 1989 publication [16]. Its application to mobile networks is also not new [17].

5.4 Requirements Analysis and Underlying Assumptions

The following is an outline of the high-level requirements for our 6G-ACE *concept* proposal protocol. A practical realization of the concept would need more study, specific functional requirements, and architectural inputs.

5.4.1 Technology drivers for 6G

The ITU-R M.2516-0 report provides pointers to the expected technology drivers for 6G [2]. For our part, we shall limit ourselves to the security-related aspects of the report. The report highlights a set of technology *key drivers* (Section 4.2 in [2]), and here we find that *Security and Trustworthiness* is explicitly mentioned.

Among others, we have that [2]:

"A future network may support more advanced system resilience for reliable operation and service provision, security to provide confidentiality, integrity and availability, privacy with self-sovereign data and safety regarding the impact to the environment."

"IMT towards 2030 and beyond should strive to support embedded end-to-end trust such that the level of information security in networks is significantly better than today's network."

None of the above applies specifically to 6G-ACE, but the need for confidentiality, integrity and privacy is part of what the 6G-ACE protocol should provide. The need for end-to-end trust is also worth noting. Section 5.8 in the M.2516-0 report highlights technologies to enhance trustworthiness [2]. Here, it is noted that:

"It is necessary to consider the extensive introduction of AI as well as the prospect of quantum computing."

5.4.2 Post-quantum cryptography (PQC)

To be future proof, our 6G-ACE concept proposal protocol should be quantum-safe. That is, the protocol must be made to support post-quantum cryptography. Clause 5.8.2 and 5.8.3 in M.2516-0 proceeds to discuss post-quantum security. It is noted that asymmetric ciphers may become compromised, and that lattice-based location-dependent identities may be used. It is also noted that one may construct physical layer security (PLS) solutions, which may enhance the resilience and robustness of the systems. The report also notes that "... post-quantum cryptographic algorithms are computationally heavy."

We might also add that the post-quantum algorithms are still fairly new, and that unexpected cryptographical breakthrough may happen. In fact, two of the post-quantum cryptography (PQC) algorithms submitted to the NIST PQC standardization have been reported to be broken. The SIKE algorithm is based on the supersingular isogeny Diffie–Hellman (SIDH) key exchange protocol. The supposedly *hard* problem of finding an isogeny (mapping) between elliptic curves proved to be far easier than anticipated [18]. Another PQC algorithm, called Rainbow was also recently reported to be broken [19]. We may therefore conclude that while it is necessary to prepare for a PQC future, it still is somewhat premature to depend on the currently proposed PQC algorithms. Thus, it may be preferable for our 6G-ACE concept protocol to be agnostic with respect to PQC. That is, to support PQC, but not require it.

5.4.3 Architectural assumptions

We shall assume very little concerning the 6G system architecture, except that we expect the generic principal entities and related trust assumptions to remain. We shall also choose to be ignorant with respect to concrete network functions within the operator domains. That is, the internal organization within the HE and SN domain is not considered.

Concerning the UE, we expect the functional division between the ME and the SIM to remain. The actual realization of the ACE-SIM is of little concern to us. However, we expect it to be like the eSIM, to be hosted on a tamper-resistant platform, and autonomous with respect to the ME.

5.4.4 History lesson impetus and requirements

From the history of the AKA protocols, in clause 5.3, we have derived the following design lessons for the 6G-ACE protocol. One may add a meta-level history lesson to "deprecate mechanisms that are not fully effective or brittle."

Lesson 1: The UE shall be the initiating entity

The connection setup, which includes *identity presentation*, is invariably initiated by the UE. The AKA protocol is triggered by the SN in response to the *identity presentation* procedure. In 5G, the SUCI procedure is technically independent of the 5G-AKA protocol, but it will always be succeeded by a 5G-AKA protocol run. Thus, it makes sense to let the identity presentation and authentication be part of the same procedure. The exception is when a temporary identifier (GUTI) is used for identity presentation. Then there already are existing security contexts.

Lesson 2: The Permanent Subscriber Identifier shall not be exposed

The SUPI must not be exposed, even to the ME or the SN. There is no functional need for ME or SN to know SUPI, as long as there is a unique reference to the UE.

Lesson 3: The temporary identifier shall be renewed for every transaction

The unique reference alluded to in the previous lesson is a temporary identifier. The 5G equivalent is the 5G-GUTI, and the lesson is in effect already captured by the 5G requirement. We advocate retaining this functionality.

Lesson 4: No UE information should be exposed over-the-air

The SUCI-mechanism in 5G will routinely expose *routing information* concerning the UE. The routing information in question is the mobile country code (MCC) and mobile network code (MNC). The SN needs to know this information in order to proceed. This is a potential weakness, albeit not a major one. It would be preferable to avoid exposing the (MCC,MNC)-tuple.

Lesson 5: Authentication may be based on symmetric key cryptography

Symmetric key cryptography, with sufficient precautions, is assumed to be quantum-safe. The challenge-response feature may therefore be retained, although with amendments.

Lesson 6: The use of a pre-shared authentication secret may be retained

The way that subscription credentials are distributed for the managed mobile networks lend themselves to support symmetric-key pre-shared secrets. The pre-shared secret used today is called K, and it is 128-bit wide. To remain quantum-safe, it ought to be ≥ 256-bit wide.

For 6G-ACE, we denote our pre-shared authentication key as the primary authentication key (PAK). It shall be present at the ACE-SIM and the HE.

Lesson 7: Explicit support for online mutual authentication

The contrived support for mutual entity authentication in 5G-AKA is a kludge. It is fully possible to start the entity authentication already in the identity presentation phase, and thus there is no real round-trip penalty to provide mutual entity authentication between the HE and the UE. An example of a protocol that connects the SUCI and 5G-AKA procedures can be found in [20].

Lesson 8: Security context establishment

A primary goal of the 4G/5G AKA protocols has been to establish security contexts and associated key hierarchies. At the core of a security context is a key-deriving anchor key and an index/reference. Security context establishment will be a primary goal for the 6G-ACE protocol.

Lesson 9: Distinct security contexts

In 5G, there are distinct security contexts for the HE and the SN domains. The 6G-ACE protocol should retain this feature.

Lesson 10: Minimize delegation from HE to SN

The trust relationships support delegation as such, but it is always a good idea to minimize the need for trust whenever possible. Thus, it would seem wise to only delegate the serving security context information.

Lesson 11: The interfaces must be fully protected

There are three parts to this lesson.

1) *The ME ↔ SIM interface*
   There is an implicit assumption in the AKA protocols that the ME-SIM interface can be trusted. We will be explicit about this and require that this interface is protected.

2) *The HE ↔ SN interface*
   It has been assumed that the HE-SN *control plane* communications channel is trustworthy. Historically, even though the standards have included support for protection, there has not actually been a firm requirement for this. We will be explicit about this and require the channel to be both confidentiality- and integrity protected.

3) *The HE and SN internal interfaces*
   The interfaces within the HE and SN networks must have provisions for protection. We require that the control plane channels to be both confidentiality- and integrity protected.

It is noted that the 6G-ACE protocol can only accommodate this requirement indirectly, through supporting the protection schemes with sessions key based on the security contexts.

Lesson 12: Support for perfect forward secrecy

It would be useful to support PFS in the derivation of the anchor keys.

Lesson 13: Quantum-safety

With precautions, it is generally assumed that symmetric key methods can be made quantum-safe. The common asymmetric cryptographic constructions are assumed to be susceptible to quantum attacks. It makes sense to avoid depending on such algorithms for the core authentication and anchor key derivation tasks. On the other hand, one does not necessarily want to depend on a PQC algorithm just yet. Thus, the logical choice is to be able to support both classical asymmetrical algorithms and PQC algorithms.

5.4.5 Performance requirements

When it comes to performance, we generally require the 6G-ACE protocol to be comparable to the SUCI + 5G-AKA combination. That is, the 6G-ACE protocol should have the same number of round-trips, comparable message sizes, and similar cryptographic performance or better. The exception will be for PQC algorithms, which are expected to be computationally heavy. That being said, the computational burden will not preclude PQC algorithms if they are needed.

5.4.6 Authentication goals

Historically, the notion of entity authentication has been somewhat unclear. What exactly does it mean that Alice has authenticated Bob? In the paper "A hierarchy of authentication specifications," Lowe gives a few definitions [21]. To our purpose, we want Alice to have fresh evidence of Bob, and vice versa.

1) Alice has fresh corroboration of Bob's identity.
2) Bob has fresh corroboration of Alice's identity.

By fresh we here mean that the evidence shall be in real-time with respect to executing the protocol round-trips. In correspondence with Lowe's "injective agreement" definition, we want a one-to-one correspondence and agreement on *data items*. That is, the agreed key material (the anchor keys, etc.) are the *data items*. The one-to-one criteria are assurance that aspects (1) and (2) above are cryptographically bounded to the same protocol run. With respect to 6G-ACE authentication between UE and HE, we shall require compliance with Lowe's injective agreement.

5.5 High-level Aspects

In this section we define the outline of the basics of the 6G-ACE protocol.

5.5.1 Principal entities

As was outlined in Section 5.2, we will refer to the principal entities as:

- **UE** - User entity (consisting of ACE-SIM and ME)
- **SN** - Serving network entity
- **HE** - Home environment entity

As for the EPS-AKA and 5G-AKA protocols, we have that the ME will derive the operational key hierarchies for 5G-ACE. Similarly, it is the SIM (here: ACE-SIM) that is the principal entity with respect to the subscriber authentication and derivation of the HE and SN security context key material. Thus, while we highlight the UE as an atomic principal entity, the reality is somewhat more complex.

5.5.2 Trust relationships

The trust relationships will be according to the outline given in Section 5.2. There are to be three main security contexts.

- **HE↔SN:** based on the roaming agreement. This security context is outside the scope of the 6G-ACE protocol.
- **HE↔UE:** the home context (HC), activated by the 6G-ACE protocol.
- **SN↔UE:** the serving context (SC), limited in scope by the HC.

The **HE↔SN** security context is required to be pre-established. That is, the *control plane* channel between the HE and the SN must be protected. How that is done is considered outside the scope of the 6G-ACE protocol.

5.5.3 Subscriber privacy to be retained and extended

The *subscriber identity confidentiality* functionality provided by the SUCI procedure is to be retained. That is, we do not propose to retain SUCI as such.

5.5.4 Anchor keys

A successful 6G-ACE protocol run shall result in the establishment of the HC and SC context. There will be one anchor key associated with each context. The anchor keys are key-deriving keys, which are used to derive the respective key hierarchies. These keys need to be at least 256-bit wide. The PAK must at least be the same size.

5.5.5 Principal entity identifiers

We need to have permanent identifiers for the principal entities. We will also need to have a roaming context identifier for the UE. We therefore define the following identifiers.

- SUPI: the primary UE identifier (same as for 5G)
- GUTI: the temporary UE identifier (same as for 5G)
- SNID: a unique SN reference/identifier
- HEID: a unique HE reference/identifier
- CID: the 6G-ACE context identifier (for the HC and SC contexts)

The structure of the SNID, HEID, and GUTI identifiers is not important to the 6G-ACE protocol. The CID is unique and constructed based on context data. It serves both as a unified context identifier and an ACE-SIM identifier.

The SUPI is privacy sensitive and shall only be known to ACE-SIM and HE. The GUTI only needs to be known to the SN and ME, but no real harm will be done if it is known to other entities.

5.5.6 Functional split between the ACE-SIM and the ME

The ACE-SIM should hold the SUPI and the permanent authentication key (PAK), and these should never be exposed to the ME. The HC anchor keys should be kept exclusively on the ACE-SIM. The SC anchor key is computed on the ACE-SIM, and then forwarded to the ME. The ME will use it to derive the serving context key hierarchy, etc. The ACE-SIM is abbreviated AS in the protocol descriptions.

5.6 The 6G-ACE Protocol Elements

In this section we give an outline of the 6G-ACE protocol. First, we need to define the pre-existing credentials and information elements (IEs).

5.6.1 Pre-existing credentials and information elements

The UE entity

The credentials and IEs in Table 5.1 are known to the AS prior to running the 6G-ACE protocol. The credentials are stored on the ACE-SIM.

The HE entity

The credentials and IEs in Table 5.2 are known to the HE prior to running the 6G-ACE protocol. The HE will have multiple (HEpub,HEpriv) key-pairs. The key-pair is identified by param, which also contains configuration information pertaining to the use of the key-pair, etc.

The HE will have roaming agreements with many networks, and so there will be multiple SN identifiers and associated security context data. Credentials and IEs associated with the HE-SN security context is not included in the

Table 5.1 ACE-SIM credentials and IEs.

IEs	Information	Shared
SUPI	Permanent subscription identifier	**No**
HEID	Permanent HE identifier	ME
PAK	Permanent Authentication Key	**No**
HEpub	HE public key	ME
param	HE public key parameters, etc.	ME

Table 5.2   HE credentials and IEs.

IEs	Information
SUPI	Permanent subscription identifier
HEID	Permanent HE identifier
PAK	Permanent Authentication Key
HEpub	HE public key
HEpriv	HE private key
param	HE public key parameters, etc.
SNID	Permanent SN identifier

table below. Suffice to require that the HE-SN has a pre-existing operational security context, with an associated protected channel. The HE-SN channel is not associated with any specific UE.

The SN entity

The credentials and IEs in Table 5.3 are known to the SN prior to running the 6G-ACE protocol. The SN does not have any a priori knowledge of the UE(s). The SN will have agreements with multiple HEs. Credentials and IEs associated with the HE-SN security context is not included in the table below. Suffice to require that the HE-SN has a pre-existing operational security context, with an associated protected channel. The HE-SN channel is not associated with any specific UE.

5.6.2 Symbols, etc.

The protocol outline is given in the Alice-Bob notation.

The symbol "→" is used to indicate message forwarding. We augment the standard notation to let the " $\twoheadrightarrow$ " indicate message forwarding over a protected channel (the HE-SN interface). The symbol "‖" indicates concatenation.

5.6.3 Cryptographic functions and IEs

Functions for symmetric and asymmetric encryption and decryption are assumed to be available. These follow the familiar $D_K(E_K(m)) = m$ notation.

Table 5.3   SN credentials and IEs.

IEs	Information
HEID	Permanent HE identifier
SNID	Permanent SN identifier
TS	Time stamp. Regularly broadcast by the SN

Table 5.4 Available cryptographic functions.

Function	Information
$\text{KDF}_K(\cdot) \to \text{Key}$	Keyed key derivation function
$\text{MAC}_K(\cdot) \to \text{mac}$	Message authentication function
$\text{prf}(\cdot) \to \text{prn}$	Pseudo-random function

For asymmetric operations, we indicate private/public keys as follows:

$$D_{Kpriv}(E_{Kpub}(m)) = m. \tag{5.9}$$

The cipher primitives are assumed to be *effective*. That is, breaking or reversing them is presumed to be *computationally infeasible*.

5.6.4 Context expiry

The context expiry condition is constructed and encoded in the Expiry IE. It is derived by the UE on the basis of the SN broadcast network time (TS) and an expiry period parameter. The UE will check the TS for soundness. The expiry period parameter is decided by HE policy, and available to the UE a priori. The resolution and encoding of the element are left for further study.

5.6.5 Challenge-response IEs

The UE and HE will challenge each other during the 6G-ACE protocol run. The challenges are pseudo-random fields. They are required to be unique and unpredictable. The responses are computed over the challenge and the CID. The inclusion of CID ensures proper context binding.

$$\text{prf}(\cdot) \to \text{CH}$$
$$\text{Res}_{\text{PAK}}(\text{CID}, \text{CH}) \to \text{RES}. \tag{5.10}$$

5.6.6 The 6G-ACE context identifier

The home security context (HC) is identified by the context identifier (CID). The CID is also used as a reference to the serving security context (SC). The CID is constructed by the HE and the UE under control of the PAK.

The pre-existing long-term HE-SN context and associated protected channels are used to forward the CID from the HE to the SN. The input arguments include the identifiers of the context participants, an expiry condition and the UE challenge.

$$\text{SIDIE} = \text{SUPI} \,||\, \text{SNID} \,||\, \text{HEID} \,||\, \text{Expiry}$$
$$\text{MAC}_{\text{PAK}}\,(\text{SIDIE}, \text{CH}_{\text{AS}}) \rightarrow \text{CID}.$$

The inclusion of the AS challenge (CH_{AS}) ensures uniqueness. The CID is an authenticated identifier, and it provides assurance of the authenticity of the AS challenge. The CID is seen as privacy sensitive as it is a unique reference to the subscriber for the duration of the primary 6G-ACE context. It should therefore not be exposed during the 6G-ACE protocol run.

5.6.7 Anchor keys

The HE-AS anchor key is denoted HAK. It is computed by the HE and the AS, and is under control of the PAK.

$$\text{KDF}_{\text{PAK}}\,(\text{SIDIE}, \text{CH}_{\text{AS}}, \text{CH}_{\text{HE}}) \rightarrow \text{HAK}. \qquad (5.11)$$

The SN-AS anchor key is denoted SAK. It is computed by the HE and the AS and is under control of the HAK.

$$\text{KDF}_{\text{HAK}}\,(\text{CID}, \text{count}) \rightarrow \text{SAK}. \qquad (5.12)$$

The counter (count) is used for SAK re-keying during the lifetime of a primary 6G-ACE context. For the 6G-ACE protocol run, count is set to a pre-defined fixed number. The re-keying will be run (logically) end-to-end between AS and HE. We have otherwise not defined the re-keying procedure.

5.6.8 Concerning the use of asymmetric cryptography

The asymmetric encryption used between the UE and HE serves only one purpose, namely, to conceal the SUPI identifier. Should the asymmetric encryption be broken, all that will be lost is the identity confidentiality of the UE. The context setup and key derivations does not in any way depend on SUPI being concealed. The asymmetric encryption could be by means of a classical asymmetric encryption algorithm or it could be a post-quantum algorithm. We have alluded to the fact that a post-quantum algorithm like the Crystals-KYBER algorithm [22] might be used. The choice of algorithm is to be encoded in the param information element. The UE, and in particular the ACE-SIM, would obviously need to support the algorithms to be used (and associated parameter, etc). We shall remain agnostic with respect to the actual cipher suite being used.

It is finally noted that the SN will not need to provide any form of support for the chosen asymmetric algorithm. This matter is entirely decided by the HE, and the HE will provide the ACE-SIM with whatever credentials needed.

5.7 An Alice-Bob Outline of the 6G-ACE Protocol

We follow the common (informal) Alice-Bob notation in our protocol outline.

5.7.1 The Alice-Bob outline

The following is an Alice-Bob like outline of the 6G-ACE protocol. For a fully defined protocol, it is strongly advised to include a protocol identifier.

One should also explicitly define error handling, and recovery mechanisms. This must be done while ensuring that privacy properties, etc., are all still fully accounted for.

The basic Alice-Bob steps are followed by a somewhat more elaborate explanation of the individual steps.

The 6G-ACE Protocol (Alice-Bob like notation)		
1:	AS→SN:	msg1(HEID,param,{SUPI,Expiry,CH_{AS},CID)$_{HEpub}$)
2:	SN→HE:	msg1(HEID,param,{SUPI,Expiry,CHAS,CID)HEpub)
3:	HE→SN:	msg2(Expiry',CID,SAK,CH_{HE},{Expiry',RES_{AS}}$_{HAK}$)
4:	SN→AS:	msg3(CH_{HE},{Expiry',GUTI}$_{SAK}$,
		{Expiry',RES_{AS}}$_{HAK}$)
5:	AS→SN:	msg4({CID,RES_{AS}}$_{SAK}$)
6:	SN→HE:	msg5(CID,RES_{HE})

We will now proceed to discuss the steps in the outline.

Step 1: The initial message
   Prior to sending **msg1**, the AS will:

- Construct/compute the Expiry, CH_{AS}, and CID.
- Encrypt the challenge data (incl. SUPI) with the HE public key.

Step 2: SN forwarding to the HE
   The SN forwards the **msg1** to the HE over the protected channel.

Step 3: HE response, forwarding to the SN
   Upon receiving the **msg2**, the HE proceeds to:

- Decrypt the challenge data. The HE then *sees* SUPI,Expiry,CH_{UE},CID.
- Verify that the public-key was intended for the specific ACE-SIM.

- Compute CID for the given SUPI (with the associated PAK), supplying HEID and SNID from the context.
- Verify that the CID is authentic (compare computed and received CID). This implies that the challenge is authentic.
- Verify the soundness of the Expiry, and potentially modify it.
- Compute a response to the challenge (RES_{AS}).
- Compute HE-AS anchor key HAK and SN-AS anchor key SAK.
- Construct and forward the **msg2** (over the protected channel).

The HE may modify the Expiry, which is why it has been denoted as Expiry'.

Step 4: SN forwarding to AS

Upon receiving **msg2**, the SN proceeds to:

- Accept the Expiry', CID and SAK, and see the CH_{HE}.
- See the encrypted data: $\{Expiry', RES_{AS}\}_{HAK}$.
- Construct a GUTI for SN-UE use.
- Use the SAK to encrypt Expiry' and GUTI.
- Construct and forward **msg3**.

Step 5: AS final forwarding to SN

Upon receiving **msg3**, the AS proceeds to:

- Receive the HE challenge: CH_{HE}.
- Compute HAK and SAK.
- Decrypt: $\{Expiry', GUTI\}_{SAK}$ and $\{Expiry', RES_{AS}\}_{HAK}$.
- Verify the received RES_{AS} (which authenticates the HE).
- Accept Expiry' (from HE) and verify Expiry' from SN.
- Accept GUTI for SN-UE use.
- Compute RES_{HE}.
- Construct and forward **msg4**.

Step 6: SN final forwarding to HE

Upon receiving **msg4**, the SN proceeds to:

- Decrypt the contents.
- Accept CID as mutually acknowledged by HE and AS.
- Accept SAK as the anchor key for the local SN-AS context.
- Construct and forward **msg5**.

Final processing at HE

Upon receiving **msg5**, the HE proceeds to verify the response.

5.8 Brief Analysis of the 6G-ACE Protocol Proposal

As explicitly expressed earlier in the paper, the 6G-ACE protocol is intended to be a concept proposal. Thus, not all details have been fully defined.

5.8.1 High-level summary

The 6G-ACE protocol has many similarities with the AKA protocol lineage, but significantly it provides online mutual entity authentication (UE-HE) and it avoids the USIM dependency. The disadvantage to avoiding USIM is of course that one then has to replace the USIM. The USIM and the associated UMTS-AKA scheme dates back to around the year 2000. Given that 6G is targeted for 2030, it should not be unreasonable to replace the USIM.

The 6G-ACE protocol provides mutual entity authentication between UE and HE. This is done by a straightforward double challenge-response scheme with MAC signed responses, and with CID as the IE that binds the exchange parts together. The primary authentication key, PAK, is the controlling key. The other high-level goal relates to provisioning of security contexts. The CID serves as a context identifier of the HC and SC contexts.

- The HC (HE-AS) context: CID,HAK
- The SC (SN-UE) context: CID,GUTI,SAK

The SC context is dependent on the HC context. There will be additional context and session keys derived according to the radio access needs.

5.8.2 Comparison with SUCI + 5G-AKA

There are a few noteworthy differences between 6G-ACE and the 5G equivalent. First, in 5G one will need to run both SUCI and 5G-AKA. This is normally done in 5G, but there is no guarantee that it is done. We also note that in 5G one does not quite have mutual entity authentication.

It would have been possible to bind SUCI and 5G-AKA together into one atomic procedure. The paper [20] provides an example of that, and one would then also have a Diffie–Hellman secret as the basis for the established contexts. This would have provided perfect forward secrecy. However, the ECIES DH secret, as derived in the SUCI procedure, is not quantum-safe. While it might be possible to make modifications, introducing PQC algorithms to the protocol, it would be a somewhat risky proposal. We are still reluctant to base the anchor key material on PQC algorithms, as these algorithms are still not considered fully mature.

It was therefore a deliberate design decision not to rely on PQC algorithms for the secrecy of the anchor keys, etc. We did accept that subscriber identity privacy be vulnerable to quantum attacks, but this is quite a different matter from permitting that the authentication and context establishment potentially be compromised.

5.8.3 Authentication properties

According to the classification in [21], the 6G-ACE protocol provides injective agreement between the HE and the UE. With respect to the SN, it will receive the session context credentials from the HE and have assurance that these have been agreed with the UE. This is acceptable provided that the HE is seen as a trusted authority (an authentication centre). Interestingly, the historical 2G/GSM name for the node that handled the authentication at the HE was the "Authentication Centre (AuC)." Thus, an assumption about the HE being a trusted authority is in line with the historical assumption.

5.8.4 Round-trip performance

The overall number of messages and round-trips in 6G-ACE is similar to the SUCI + 5G-AKA protocols. The 6G-ACE protocol permits the SN to activate the SC after have handled step 4.

5.8.5 Adherence to the history lesson requirements

We have listed a set of design lessons in Section 5.3. Most lessons have trivially been adhered to. However, there are a few deviations. Some of these are considered outside the scope of 6G-ACE, but there are also some goals that are not met or not fully met. Those lessons are discussed below.

Lesson 5: No UE information should be exposed over-the-air
This requirement is mostly fulfilled, but we note that the HEID and param are exposed over-the-air. This is comparable to the 5G SUCI case. Depending on circumstances, this may be a liability. However, we note that the "IMT-2030 and Beyond" technical trends report highlights the possibility to use federated learning to enhance "RAN privacy" (Section 5.8.1) and physical layer security (Section 5.8.3) to provide adaptive security for the radio connection [2]. These technologies would be well suited for solving this problem, and we advise investigating this further.

Lesson 11: The interfaces must be fully protected
This is an important requirement, but it is also considered outside the scope of the 6G-ACE protocol to provide this.

Lesson 12: Support for perfect forward secrecy
This goal is not achieved. It could have been achieved by including a DH exchange, but after considerations this was deliberately not done. The PFS property, while desirable, is not, in practice, very important for access security. The added cost seems not worthwhile. Furthermore, classical DH methods are susceptible to quantum-attacks and the quantum-safe algorithms are still somewhat immature.

Concerning *Lesson 13: Quantum-safety*, the 6G-ACE protocol has been designed to permit use of PQC protocols. 6G-ACE uses asymmetric primitives for subscriber identity concealment. In our design, we have remained agnostic to the choice of asymmetric algorithm, and an algorithm like Crystals-KYBER may be used. The other parts of 6G-ACE depend on primitives that may all be realized by MAC/hash algorithms or symmetric ciphers. These can generally be made quantum-safe by doubling the key-size. Thus, 6G-ACE is considered a quantum-safe design.

5.9 On the Completeness of the 6G-ACE Concept Proposal

5.9.1 Key derivations and key hierarchies

We have deliberately not investigated how to construct the key hierarchies that needs to be derived. In fact, we do not consider key derivations to be the responsibility of the 6G-ACE protocol and procedures, except for derivation of the anchor keys and keys used directly by the protocol.

The actual key hierarchies will also need to be designed for the 6G radio access network, which are not yet defined.

5.9.2 Error handling

It has already been mentioned that the 6G-ACE concept proposal does not include any error handling. It is important that a fully-fledged protocol also convey error information, although great care must be taken to avoid unnecessary information leakage. This is particularly urgent for the identity privacy properties, as leaked information may give rise to emergent properties and patterns.

5.9.3 Context mapping and backwards compatibility

We have not investigated context mapping from 6G to 5G, but this may be both desirable and feasible.

Mappings from 5G *may* be acceptable as such, but it will introduce dependencies and complexities that may potentially weaken the 6G security. We shall therefore advocate against mapping from previous generations to 6G. Other requests for backwards compatibility should generally be turned down, and only after intense scrutiny should it be permitted (if at all).

5.9.4 Future work

Needless to say, but the 6G-ACE concept proposal will need to be amended and updated should it ever become a candidate for actual adoption for 6G. Furthermore, one should provide a formal model of the 6G-ACE protocol.

The value of "proofs" may sometimes be overstated, but even trivial proofs will have some value, and the construction of a validation model may also lead to new insights concerning the protocol in question.

It may be necessary to provide ultra-lightweight solutions for machine-to-machine (m2m) communications. This would target devices that have extreme power constraints. This may entail using dedicated light-weight ciphers, like the Ascon. Ascon was chosen by NIST as a "Lightweight cryptography" standard in February 2023, and should be suitable for lightweight m2m/IoT uses [23].

5.9.5 Cryptographic safety and agility

There should be a critical review of which cryptographic primitives to be used. One should aim to ensure that there is a level of flexibility with regard to managing the cryptographic primitives. This could include measure that may mitigate quantum attacks and may range from extending the key length for symmetric methods to replacing standard asymmetric methods with quantum-safe ones.

5.10 Summary

We have provided a brief account of the history of the Authentication and Key Agreement protocols of the 3GPP-based systems, dating back to the pre-1990 design of the GSM-AKA protocol. The successive generations, with approximately a decade between them, have improved and extended on the functionality of the previous generation. Design-wise, there has been an adapt and amend approach, which has suited the mobile system evolution well. However, this approach also makes it harder to avoid backwards compatibility constraints and to come up with clean designs. From a security perspective, complexity is highly undesirable.

Protocol design is difficult, and subtle design flaws may hide in complex schemes. And, even for flawless designs, complexity is a problem in that implementations must also be in full compliance with the design.

To demonstrate that one can come up with simpler designs that still provide the necessary functionality, we have designed a concept proposal for a 6G protocol.

Given that the 6G system architecture is not decided yet, we had to restrict ourselves to what we see as the invariant parts of the mobile systems. That is, the entity types and the trust relations between the principals. For our purpose, we have the home environment (HE) entity type and the serving network (SN) entity type. For the user entity (UE), we have an explicit split between the mobile equipment and the subscriber identity module.

The 6G-ACE concept proposal protocol breaks with existing AKA designs in some ways, in order to simply the design. The overall structure of 6G-ACE is in line with the basic scheme of "SUCI + 5G-AKA'," and 6G-ACE also provides SUCI-like functionality. The security context and anchor key concepts are retained.

Given the "concept proposal" theme and the fact that there is yet no 6G architecture, we obviously do not expect the 6G-ACE protocol to be a complete protocol proposal for 6G.

5.11 Conclusion

The 3GPP-based authentication and key agreement protocols cover several mobile system generations (2G–5G). There is a clear linage and evolutionary development. While that has served the systems well, the inherited features also contain features that make the AKA protocols more complex than they would have had to be. There are also a few "omissions" that can be traced back to design decision made during the 3G design. Our 6G-ACE protocol has retained features and requirements from the 3GPP AKA linage that seems future-proof. There is no dramatic break from history in the 6G-ACE protocol, but we have avoided backwards compatibility when necessary and this has allowed for a simplified and improved design.

To conclude, it is hoped that our concept proposal can inspire further research in this area, and ultimately that our proposal can provide an impetus for the future standard work in this area.

References

[1] International Telecommunication Union (ITU), "IMT Vision – Framework and overall objectives of the future development of IMT for 2020 and beyond". Recommendation M.2083-0, ITU-R, 09 2015.

[2] International Telecommunication Union (ITU), "Future technology trends of terrestrial International Mobile Telecommunications systems towards 2030 and beyond". Report M.2516-0, ITU, 11 2022.

[3] Danny Dolev, and Andrew Yao. On the security of public key protocols. IEEE Transactions on information theory 29.2 (1983): 198-208.

[4] Nordic Mobile Telephone group. System Description. Number NMT Doc 450-1 1995.

[5] 3GPP. TS 43.020 "Security related network functions", Rel.17.0, 03 2022.

[6] 3GPP. TS 33.102 "3G security; Security architecture", Rel.17.0, 03 2022.

[7] Geir M. Køien. An introduction to access security in UMTS. IEEE Wireless Communications, 11(1):8–18, 02 2004.

[8] Valtteri Niemi and Kaisa Nyberg. UMTS Security. John Wiley & Sons, 2003.

[9] 3GPP. TS 33.401 "3GPP System Architecture Evolution (SAE); Security architecture", Rel.17.3, 09 2022.

[10] Dan Forsberg, Günther Horn, Wolf-Dietrich Moeller, and Valtteri Niemi. LTE Security (2nd edition). John Wiley & Sons, 2012.

[11] 3GPP. TS 33.501 "Security architecture and procedures for 5G System", Rel.18.1, 03 2023.

[12] Anand R Prasad, Sivabalan Arumugam, B Sheeba, and Alf Zugenmaier. 3GPP 5G security. Journal of ICT Standardization, 6(1):137–158, 2018.

[13] Standards for Efficient Cryptography Group (SECG). SECG SEC 1: Recommended Elliptic Curve Cryptography, Version 2.0. Technical report, SECG, 2009.

[14] Standards for Efficient Cryptography Group (SECG). SECG SEC 2: Recommended Elliptic Curve Domain Parameters, Version 2.0. Technical report, SECG, 2010.

[15] GSM Association. eSIM Architecture Specification; Version 3.0. Permanent Reference Document SGP.21, GSMA, 03 2022.

[16] Christoph G Günther. An identity-based key-exchange protocol. In *Workshop on the Theory and Application of Cryptographic Techniques*, pages 29–37. Springer, 1989.

[17] DongGook Park, Colin Boyd, and Sang-Jae Moon. Forward secrecy and its application to future mobile communications security. In *International Workshop on Public Key Cryptography*, pages 433–445. Springer, 2000.

[18] Wouter Castryck and Thomas Decru. An efficient key recovery attack on SIDH. Annual International Conference on the Theory and Applications of Cryptographic Techniques. Springer Nature Switzerland, 2023.

[19] Ward Beullens."Breaking Rainbow takes a weekend on a laptop." In *Proceedings of CRYPTO 2022*, Santa Barbara, CA, USA, August pages 464–479, Springer 2022

[20] Geir M Køien. The SUCI-AKA Authentication Protocol for 5G Systems. In *Norwegian Information Security Conference*, no.3, 2020.

[21] Gavin Lowe. A hierarchy of authentication specifications. In *Proceedings 10th Computer Security Foundations Workshop*, pages 31–43. IEEE, 1997.

[22] Joppe Bos, et al. "CRYSTALS-Kyber: a CCA-secure module-lattice-based KEM." In *Proceeding of IEEE European Symposium on Security and Privacy 2018 (Euro S&P)*. IEEE, 2018.

[23] Christoph Dobraunig, et al. Ascon v1. 2: Lightweight authenticated encryption and hashing. Journal of Cryptology, 34:1–42, 2021.

Biography

Geir M. Køien started his professional career while working for LM Ericsson Norway, where he worked with software testing of NMT systems. He has worked both in industry and academia, and while at Telenor R&D he was the Telenor delegate to 3GPP workgroup SA3 (Security). He received his Ph.D. degree from Aalborg University and is currently a professor of Cybersecurity at the University of South-Eastern Norway.

6

ICT Applications in Health Monitoring

Torsten Wipiejewski, Yaxin Xu, and Walter Weigel

Huawei European Research Institute, Belgium

Abstract

With the quick development of information and communication technology, the healthcare industry profits quite a lot using ICT as an enabling technology. In this chapter the directions of health research at Huawei for vital signs monitoring will be introduced. The convenience and reliability of medical technology applications can only be guaranteed with the appropriate implementation of efficient interfacing electronics and novel sensor solutions (e.g., MEMS, optical, micro-fluidics, etc.) especially in consumer products. Together with the development of 5G and 6G wireless communications, the ICT is becoming a fundamental driver for this ecosystem of connected health devices.

Keywords: ICT; Health Care, Artificial Intelligence, MEMS etc.

6.1 Introduction

Health monitoring is an area where much attention is paid to because of the hope that it can significantly reduce the cost in the health systems by identifying sicknesses as early as possible. It is usually performed at various levels with different time intervals as depicted in Figure 6.1. People who are not actually sick would typically go to see a doctor occasionally for a general check-up. The frequency varies of course, but it is typically every one to two years. The monitoring equipment are medical-grade laboratory style equipment which provide high accuracy. Body fluids and body functions are checked and compared to normal average values and monitored over the time from visit to visit. Thus, the health status is only checked occasionally with long time periods in between.

Time interval	Location/user	Example	Requirement
Occasionally	Clinic, Medical experts	Health check-up Specific inquiry	Accuracy
Periodically	Home usage by user	Weight Blood pressure Glucose (for diabetes patients)	Low cost Easy to use Small size
Constantly	Close to body by user	Heart rate Sleep Activity (e.g. step counting)	Wearable Battery powered Connectivity

Figure 6.1 Levels of typical health monitoring for actually healthy people.

The idea is that more frequent health monitoring can be performed by people themselves at home. Typical examples are watching the weight or measuring blood pressure. The latter is mostly done by elderly people. Individuals who suffer from chronic diseases of course would monitor other body parameters such as blood glucose, but it is not common for people without any indication of a disease. As people are supposed to use the health monitoring device at home, the cost of the devices must be affordable and the handling easy enough for non-trained users. This automatically limits the scope of health monitoring that can be performed.

A relatively new way of health monitoring is provided by personal devices such as smart phones, smart watches, and smart bands. Figure 6.2 shows products offered by Huawei.

Micro technology and miniaturization of components enable the integration of various sensors. The high computing power of the smart devices can be used to analyze the raw data from the sensor elements and transform them into useful information [1].

Smart Phone   **Smart Watch**   **Smart Band**

Figure 6.2 Examples of smart devices for personal usage. Micro technology enables the integration of various sensors.

Optical sensors and MEMS-based motion sensors are used for the monitoring of heart rate, blood oxygen levels, sleep condition, and others. Step counting functions can help users monitor their own behavior and can motivate them to perform more physical activities. Many functions are useful for health monitoring and fitness monitoring alike.

The fundamental advantage of these smart devices is their small size that enables users to carry them at all times. They can be attached to the body like a watch and become "wearable." Constant monitoring of body functions is now possible. The devices are battery powered. Thus, energy consumption should be minimized and an easy battery charging solution must be provided. This is either done with a charging cable or by wireless charging methods.

Although the devices have a build-in intelligence to analyze sensor signals and display information to the user, they also provide connectivity for further analysis and data storage (Figure 6.3). Typically, the devices can be connected with a health application software through an Internet connection. With data security protocols the health monitoring data can also be shared with medical experts for a more thorough diagnosis. The data of many users can be analyzed by statistical methods and by using artificial intelligence. A "big data" system can provide additional information and predictions about the health status. This can assist in the diagnosis and potentially in prevention of therapy.

The possibility of monitoring patient health data in a continuous way over a longer period of time provides a new tool for doctors which was not easily

Figure 6.3 Heart rate data monitored by the smart watch can be transferred to a smart phone for further analysis and data storage. Additional evaluation of the data can be performed in data centers using "big data analytics."

available in the past and will provide a new quality of health diagnostics. For example, some people suffer from heart rhythm issues that occur only occasionally. It is difficult for the doctor to evaluate the issue if the patient can only be monitored during a short visit to the clinic, because the issue might not occur at all during this short time period. Thus, a constant monitoring over longer time periods can provide more valuable data.

6.2 Mobile Health Solutions

Wearable devices are convenient as they can be attached to the body by straps or similar solutions. The position of the wearable is determined by the functional requirement and the ease of use (Figure 6.4). The most popular spot is the wrist which has been used for normal watches for a long time. It also enables access to blood vessels for optical monitoring of the blood signal using a photo sensor combined with light from an LED. This method is called photo plethysmo gram (PPG). Sometimes, several LEDs are employed for improved accuracy. Another important aspect for wearable devices is the appearance. The device must also fulfill fashion and aesthetic requirements to be accepted by the user.

Figure 6.5 shows possible positions for wearable devices. Very common are ear phones (ear buds) that people use to play music or listen to voice signals. The position inside the ear enables the integration of sensors that can detect blood signals or body temperature.

Other positions on the body include the waist using a belt or the ankle or foot in combination with smart shoes. These can help to detect the walking or running motion in detail. Besides potential health monitoring it can also assist people in improving the motion for running or other sports activities.

Choice of Position for Wearable Devices:

➤ **Easy and convenient to access**

➤ **Medical requirements (PPG sensor)**

➤ **Good looking**

Figure 6.4   Conditions for the position of wearable devices.

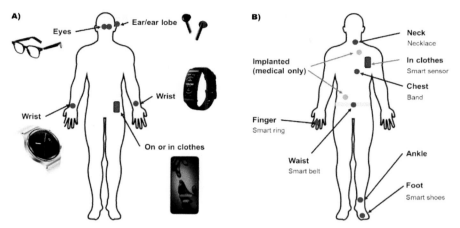

Figure 6.5 Typical positions on the body for wearable devices. (A) mostly used today, (B) additional positions in the future.

The chest position requires a belt around the body or a patch with adhesive properties. It is mainly used if the position of the sensor must be close to the heart or lung. Necklaces and rings could also host sensing functions, but these positions are typically considered for decorative elements. Therefore, good appearance is of high importance.

6.3 Hardware Requirements

The integration of more and more functions and the drive to higher performance, faster speed, and more connectivity pushes the development of electronic devices. Figure 6.6 depicts some key requirements for smart devices.

The small size is most important to accommodate the various devices including displays and batteries that need as much space as possible for best user experience. Low power consumption of all components is necessary for a long battery operating time and the ability to provide enough cooling for the devices. High reliability and low cost are other important factors for successful products in the market. A way to fulfill these requirements is integration. Monolithic integration is the backbone for CMOS circuits going to smaller transistor feature sizes and larger number of elements integrated on a single chip. The well-known Moore's law describes the evolution of monolithic integration over many decades and is still the guideline for chip development today.

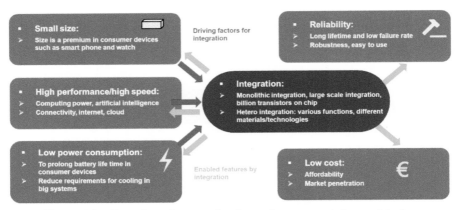

Figure 6.6   General requirements for electronic components in smart devices.

Packaging and the heterogeneous integration of various chips and elements are other key technologies to achieve the small size, high performance, and low-cost target of smart devices, in particular wearable devices. Heterogeneous integration offers the possibility to combine elements (chips) that require different materials and/or different manufacturing processes. For example, electronic circuits are mostly based on silicon chips whereas light emitting devices such as LEDs are based on III–V semiconductors such as GaN or InAlGaAs. A monolithic integration of these different materials onto a common substrate has been attempted for a long time, but the performance is typically sacrificed by the combination of very different processing steps. Thus, a heterogeneous integration of two chips, each with optimized material selection and processing steps, can be the preferred solution.

Even if the performance is not compromised, monolithic integration is not always the best choice. In many sensor applications the size of the sensing element is determined by a specific application and can be relatively large. The cost of the electronics CMOS manufacturing process, however, scales with area. Thus, the electronics circuit becomes unnecessarily expensive if a lot of the wafer area is used up for the sensing element. Similar conditions apply to the integration of optical elements with electronics. Figure 6.7 shows a comparison of monolithic and heterogenous integration for the example of a MEMS element with a CMOS circuit.

The question of which integration is more advantageous is complex and must be analyzed for specific cases. In general, the heterogeneous integration tends to be more flexible. It allows a wide selection of technology steps

	CMOS
MEMS	

MEMS CMOS	MEMS CMOS

TOPIC	Monolithic	Hetero Integration
Area of electronic chip/part	Big chip area including MEMS	Small, just function related
Assembly process	Single chip, simple	Die to die bonding or co-packaging
Testing	Chip with entire function	MEMS, electronic chip separately
Impact of yield	Cost of chip with full process chain	Cost of faulty chip only, less complexity

Figure 6.7 Comparison of monolithic versus heterogeneous integration of a MEMS technology chip and a CMOS electronic chip.

and materials. It also provides more flexibility when combining functions of different development cycles such as CMOS circuits and photonic elements.

Several options exist for the heterogenous integration of MEMS sensor components with electronics CMOS chips. Many sensor chips require a hermetic sealing to protect the moving MEMS element from dust and moisture. It also provides constant ambient conditions whereas the signal might otherwise vary too much. Two packaging sealing options are depicted in Figure 6.8.

The MEMS sensor and the CMOS chip can be co-packaged on a common sub-mount. The sub-mount provides the electrical connection and mechanical

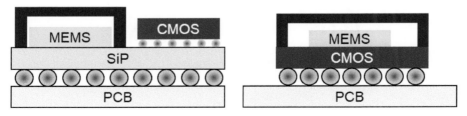

Figure 6.8 Packaging options for MEMS sensor elements.

stability. It can also host other elements such as coils or larger capacitors. The components can be tested before mounting on a larger PCB. In another approach, the MEMS sensor is directly mounted on top of a CMOS chip. The electrical connections in CMOS chip for MEMS Sensors are shorter which reduces the parasitic effects. However, the MEMS element is also heated by the underlying CMOS chip, which must be considered in the design.

6.4 Textile Integration

Instead of using bands to attach wearable devices to the body they can also be integrated in textiles as illustrated in Figure 6.9.

Although textile integrated electronics was already demonstrated many years ago, the size of the market is still very limited. Part of the reason is technical challenges, because the packaging of the electronic devices must be very rugged to withstand typical washing cycles.

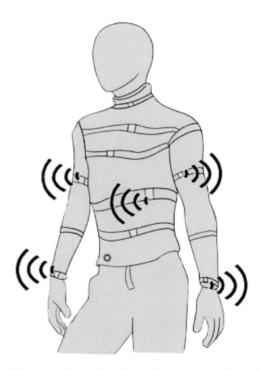

Figure 6.9   Integration of health monitoring sensors in textiles.

Difficult technology	**Difficult value proposition**
➢ **Washing cycles**	➢ **Attachable devices provide the same or better performance**
➢ **Stretching**	
➢ **Comfortable feeling**	➢ **High cost electronics compared to low cost textiles**

Figure 6.10   Challenges for textile integrated sensing devices.

It is not easy for electronics to survive hot water with detergent many times. Typically, it is also a requirement that the components or at least the electrical connections are stretchable like the fabric and they must provide a comfortable feeling for users (Figure 6.10).

Another challenge for textile integrated electronic devices is the value proposition, because most functions can also be provided by discrete devices with the same or even better performance. These discrete devices are attachable and can be removed after usage and before washing. The biggest advantage of the textile integration might arise in cases when a larger area of the body needs a certain treatment such as light therapy or when the whole body is a sensing area for example in gaming applications.

Another challenge for textile integrated electronic devices is on the economic side. Even low-cost electronics devices tend to be more costly than typical fabrics. Therefore, the current usage focuses on special applications where the integration can provide great benefit and the cost of the clothes is not very sensitive.

6.5 Future Diagnostics

So far, health monitoring focuses on non-invasive sensing technologies that can provide information on some of the basic body functions. Since blood is transported to all parts of the body it provides a lot of valuable information on the health status of a person. In classical medicine diagnostic blood samples are routinely characterized. The methods are mostly based on chemical analysis. For consumer applications it is more challenging to take blood samples because of potential complications and the chance to get infections by the intrusion of objects into the skin. Micro needles and other extraction methods have been explored to access samples of body fluids with only minimum intrusion. The methods are normally referred to minimum invasive. Small chemical analysis reactors with micro fluid channels and chambers are employed for a small size, low cost chemical analysis.

For convenience and safety, non-invasive sensing technologies are still preferred. Most of them are based on optical detection technologies. The whole optical spectrum from visible to far infrared could be considered. Especially, the mid-infrared spectral range can provide characteristic spectral absorption patterns that can be utilized for the detection and identification of certain molecules. However, the light penetration into the skin must also be considered. Most of the spectrum is blocked by the absorption of the skin, mainly due to water molecules. Thus, the detection should measure the spectrum where the molecule of interest shows a characteristic spectral signature and the skin provides sufficient transparency for the particular wavelength range. Sometimes, molecules of interest are not detected in the blood, but in upper layers of the skin in the interstitial fluid.

An important example for a non-invasive optical detection is glucose. Monitoring the blood glucose level is crucial for diabetes patients. Various methods and wavelength ranges have been investigated such as Raman spectroscopy in the near infrared or absorption spectroscopy in the mid infrared. A big challenge for all these technologies is that, from a medical point of view, a very low glucose level in the blood is of high interest, but of course it is difficult to measure something that is almost not there. Thus, the detection limit must be very low and the sensor very sensitive.

It is expected that the optical detection of other substances such as lactate or alcohol in the blood or even gases in the breath will play an increasing role in the future. Although it has not been as widely used as the analysis of substance in the blood, breath analysis can also provide a lot of information on the health status of a person. In addition, exhaled breath can characterize body functions, in particular the digestive system (Figure 6.11).

Besides CO_2-enriched air there are dozens of volatile organic compounds in the exhaled breath. Each one can be an indicator for certain processes in the body and be an early warning indicator for any malfunction or disease. Dogs with their very sensitive nose have been successfully employed for breath detection for many years. A more technical approach using the very sensitive optical detection of specific gas molecules can be considered to be used more widely in the future.

Many sensor elements are already available to users such as step counting, heart rate monitoring, ECG, blood oxygen saturation, and others. The smart watch or the smart band are the dominant platform for the integration of the sensor elements, because the position at the wrist is convenient for the user and enables the optical access to blood vessels. It is expected that in the future

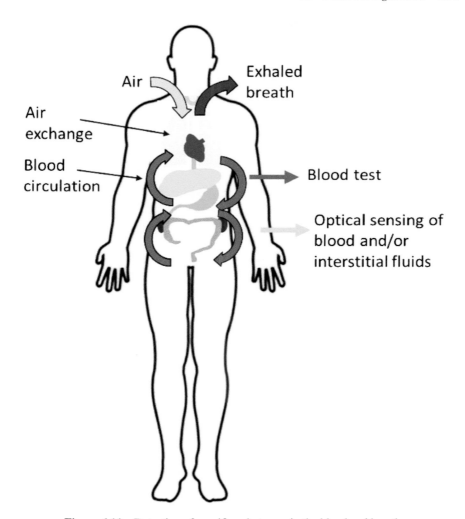

Figure 6.11 Detection of specific substances in the blood and breath.

the combination of various sensors will become more important to improve the accuracy of the health monitoring system and to enable further diagnoses. Besides the wrist position, other potential positions on the body will be fitted with sensor elements and all sensors will be connected in a network to enable a "sensor fusion" scheme as shown in Figure 6.12.

The combination of various sensor elements will help users to monitor their health status and to receive early warning signals in case of potential health threats. It will also assist health professionals in evaluating the health status of a patient, because the data is available over a longer period of time or even continuously.

For convenience and ease-of-use, the various sensor elements on the body will be connected by wireless signals. The wireless system should provide

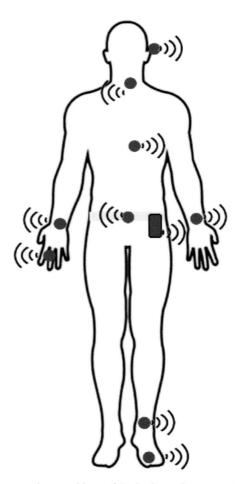

Figure 6.12   Sensors on various positions of the body can be connected by wireless communication technology. The data from the different sensors can be combined to provide additional information from the sensor fusion.

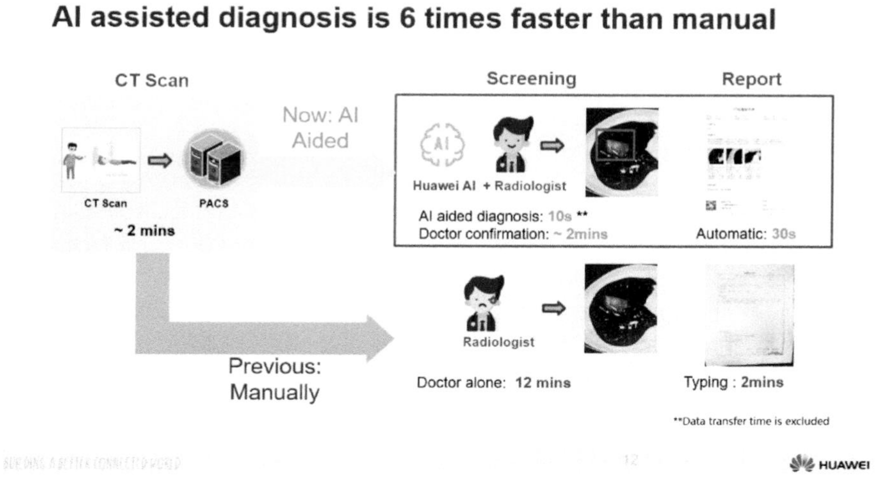

AI assisted diagnosis is 6 times faster than manual

Figure 6.13 AI-assisted diagnosis.

a high quality-of-service with high reliability combined with low power consumption for the battery-powered operation of the sensors. In some cases, the power consumption of the sensor elements might be low enough to utilize energy harvesting instead of a long-lasting battery for sensor operation and connectivity. Energy harvesting from light, mechanical pressure, or movement as well as thermal energy are typical options, yet due to the efficiency only for ultra-low energy devices.

Although the data rate for sensing signals is typically low in comparison to the high-speed signals required for voice and video type data transmission, the latency must still be considered, especially in sensor fusion schemes when the signals of several sensors must be synchronized.

The connection of health monitoring functions from individual users to medical experts will also improve the efficiency of the healthcare system, as people can stay at home while they are being examined by the medical expert remotely. That will be beneficial especially in rural areas where the distance between medical expert and patients could be quite large.

Finally the advances in AI-research (shown in Figure 6.13) will enable an automated analysis in order to support medical experts and allow fast warnings to risk patients, e.g., when suffering from chronic diseases. A first example (yet related to scan pictures, not wearables) was done in a research project for Covid patients in China and delivered a six-time improvement for the diagnosis time.

6.6 Conclusion

Information and communication technology (ICT) has been playing a very important role in the advancement of medical diagnostics and treatment for many years. Highly sophisticated equipment in hospitals and doctors' clinics provide the basis for the healthcare system. A limited number of personal medical devices such as blood pressure measurement equipment are used by some people at their home. The advancement of microelectronics and the widespread usage of smart devices such as smart watches and smart bands create another opportunity for health applications now. Microelectronics and the hybrid integration of components in very small size packages enables the integration of various sensor elements in the small size of wearable devices. These are easy and convenient to use and can be worn constantly. Therefore, these devices can provide additional long-term health related data that cannot be obtained otherwise in a convenient way for the user. The long-term data is a new valuable information and can help medical experts in the diagnosis of health issues and can provide an early warning platform for users regarding potential health risks. The miniaturized and user-friendly electronic health monitoring devices work with telemedicine services to form a closed-loop system that supports continuous monitoring, early warning, and medical consultation. The user will have a continuous, convenient, and efficient healthcare experience.

Wireless technology is essential for personal health monitoring. Wireless data communication provides the connection between sensing devices and other personal devices such as a smart phone and are part of the research for 6G. This is important in order to analyze the health monitoring data, store the data and display the information in a convenient way. Future wireless communication technology will further improve the connectivity of multiple sensors and enable a more connected and more intelligent world.

References

[1] Ramjee Prasad, Anand Raghawa Prasad, Albena Mihovska, and Nidhi, "6G Enabling Technologies New Dimensions to Wireless Communication", ISBN 9788770227742, January 31, 2023 by River Publishers

Biography

Walter Weigel graduated from the Technische Universität München, Germany, with the Diplom-Degree in electrical engineering in 1984 and with the Ph.D. degree (thesis about speech recognition) in 1990. From 1984 to 1991, he was an assistant at the Lehrstuhl für Datenverarbeitung (Institute for Data Processing) at the Technische Universität München.

Dr. Weigel was, from 1st April 2015 to 30th August 2022, VP and CSO of the European Research Institute of Huawei with headquarter in Leuven, Belgium and since 1st September 2023 has been a consultant for Huawei Technologies.

He was, from September 2006 to July 2011, the Director General of the European Telecommunication Standards Institute ETSI. Between February 1991 and February 2015, he held several positions within Siemens AG, including VP of External Co-operations and Head of Standardization in Corporate Technology, VP of the Research and Concepts Department of the Mobile Networks business unit, as well as Head of the business segment Video Processing for the semiconductor business unit (today Infineon).

He was a member of the Innovations dialog of the German Government, of the BDVA (Big Data Value Association) Board of Directors, of the Senate of Acatech (German academy of technical sciences), of the Board of 5GAA (5G Automotive Association), of the Digital Manufacturing Executive Council of Digital Europe, and of the Advisory Committee of CTIF Global Capsule (CGC) of Aarhus University.

He is a member of the IEEE-SA Board of Governors and lecturer and honorary professor at Technische Universität München.

Torsten Wipiejewski is the manager of the Central Hardware and Engineering Research Institute in the Huawei ERI (European Research Institute).

Yaxin Xu is an ERI in charge of the Hardware Research of the Huawei Consumer Business Group.

7

Key Issues in NOMA from the 6G Perspective

Saurabh Srivastava, Rampravesh Kumar, and Sanjay Kumar

Department of E&TC, K. K. Wagh Institute of Engineering Education and Research, Nashik, Maharashtra, India.

Abstract

The deployment of the fifth-generation (5G) network has already begun, offering significant advancements compared to existing systems. However, solely relying on 5G may not be adequate to fulfill the increasing future requirements. As wireless communication expands beyond connecting humans to connecting machines, there will be emerging use cases that require exceptional performance. This has led to the development of the sixth-generation (6G) network, which aims to fulfill unprecedented requirements and expectations that 5G cannot fully satisfy. The non-orthogonal multiple access (NOMA) schemes have garnered substantial attention from the perspective of 5G. However, for its implementation in 6G, NOMA needs further enhancements. Several challenges need to be addressed to improve NOMA's performance. These challenges include tackling imperfect successive interference cancellation (SIC), managing interference issues, handling imperfect channel state information, optimizing signature design, reducing receiver complexity, ensuring physical layer security, optimizing power allocation, and devising efficient user-pairing strategies. Additionally, the integration of NOMA with 6G heterogeneous networks is crucial. By addressing these challenges, NOMA can enhance its performance and leverage its maximum potential benefits to meet the requirements of 6G. This chapter focuses on exploring various key research issues that need to be tackled for the performance enhancement of NOMA regarding 6G.

Keywords: NOMA, SIC, CSI, Signature, Physical Layer Security, Power Allocation, User-pairing, Heterogeneous-Network Integration

7.1 Introduction

With the on-going deployment of fifth-generation (5G) networks, the key question remains "What next?" The sixth-generation (6G) networks are on the horizon. 6G is expected to offer a highly dense network with greater reliability, higher user rates, ultra-massive connection, and extremely low network latency. Spectral access in terahertz (THz) bands, edge-computing, non-orthogonal multiple access (NOMA), quantum computing (QC), machine learning (ML), artificial intelligence (AI), and deep learning (DL) are among the cutting-edge approaches and technologies proposed toward 6G development.

Based on statistical predictions, according to projections, the quantity of Internet-connected devices is expected to experience substantial growth, with projections indicating a rise from 10.07 billion in 2021 to 25.44 billion by 2030 [1]. Currently, various sectors such as energy, water supply and waste management, transportation and storage, gas, steam and A/C, retail and wholesale, and government have already deployed over 100 million networked Internet of Things (IoT) devices [1]. In the consumer segment, which includes Internet and multimedia applications, this number is expected to exceed eight billion by 2030. Additionally, emerging use cases like IT infrastructure, smart grids, asset tracking and monitoring, and autonomous vehicles will require over one billion interconnected IoT devices [1].

In a conventional 5G communication infrastructure, it is feasible to accommodate a maximum of 50,000 IoT or narrowband IoT (NB-IoT) devices per cell [2]. However, to facilitate massive access from the perspective of beyond 5G (B5G) or 6G communication technologies, stronger along with more resilient network infrastructure is necessary. By the year 2030, international standardization agencies are anticipated to have finalized the specifications for 6G networks [3]. Researchers in [4] have further expanded the vision of 5G networks to encompass visionary technologies for next-generation wireless systems, setting the stage for 6G advancements.

As 6G technology advances, its primary emphasis will be on the THz frequency range, known for its extensive bandwidth capabilities. Interacting effectively at these frequencies will pose fresh challenges that demand efficient solutions. Ensuring a harmonious combination of spectrum performance, coverage, and power efficiency will be of utmost importance when designing devices that operate within these emerging frequency spectrums. Consequently, the development of a contemporary air interface will prioritize the consideration of single-carrier systems.

However, one drawback of operating in higher frequency ranges is the rapid attenuation of the signal during transmission, which limits its range to a few hundred meters. The potential of 6G wireless communication networks lies in their ability to provide widespread, secure, and instant wireless connectivity for both humans and computers, supporting a multitude of intelligent applications. Recent advancements in machine learning have opened doors to various new technologies such as autonomous vehicles and voice-activated assistants. The integration of AI into wireless algorithms, such as for channel estimates, channel state information (CSI) feedback, and decoding, has the potential to revolutionize the direction of these algorithms.

7.2 Fundamentals of NOMA

NOMA, a multiple access scheme, has garnered significant attention in the context of 5G. However, for its implementation in 6G, further enhancements are needed. NOMA enables several users to access the same resource elements, irrespective of time, frequency, or code, making it a prominent area of research in both academia and industry for 5G and beyond. The non-orthogonality inherent in NOMA offers an improved rate region. This is commonly achieved through transmitter superposition coding and receiver successive interference cancellation, although alternative techniques may also be employed. Through the utilization of capacity-achieving schemes, NOMA can enhance spectral efficiency and user fairness in the downlink. In the uplink, NOMA facilitates simultaneous access to the same wireless resources by multiple users, thereby supporting more connections and reducing latency. Consequently, NOMA initially emerged as a promising solution for next-generation systems.

Broadly, NOMA may be categorized depending on the process it utilizes to differentiate between the users. Thus, NOMA can be generally categorized into two main groups. The first is power-domain NOMA (PD-NOMA), where users are multiplexed and distinguished based on power levels. Further, the second is CD-NOMA where users are differentiated with non-orthogonal user signatures. In PD-NOMA, each user is assigned a unique power level for uplink transmission, and during the reception at the base station (BS), superposition coding (SC) is utilized. In the downlink, the BS constructs the SC signal and transmits it to the users.

7.3 NOMA Challenges

There are several challenges faced by NOMA such as imperfect SIC cancellation, interference issues, availability of imperfect channel state information, signature design, high receiver complexity, physical layer security issues, power allocation, and user-pairing strategies, followed by NOMA integration with 6G heterogeneous networks. Addressing these issues will improve NOMA's performance to harness its maximum potential benefits toward meeting 6G requirements. This chapter discusses various key research issues to be addressed to improve NOMA performance. Of the two NOMA classes described in the previous section, PD-NOMA is given an edge over the code domain NOMA (CD-NOMA) by the academicians owing to the simplicity of SC and SIC operations.

Mathematically, the superposed signal x_s for a two-user PD-NOMA network may be represented as:

$$x_s = g_1\sqrt{aP}x_1 + g_2\sqrt{(1-a)P}x_2 + n_{\text{channel}}, \tag{7.1}$$

where, $g_i, x_i; i = 1, 2$ respectively represent the channel gains and the messages of the users, P denotes the BS power constraint, and n_{channel} represents the additive white Gaussian noise (AWGN), that corrupts the superposed signal. The individual user messages need to be extracted from the superposed signal using the process of SIC.

7.4 Imperfect SIC

A significant portion of the existing literature focusing on PD-NOMA assumes that the SIC receiver has complete cancellation. However, in practical scenarios, achieving error-free by subtraction of the decoded signal from the received signal poses challenges. The discussion presented here applies to both the downlink and the uplink, although the analysis is specifically performed for the downlink scenario. Consider a system model consisting of a single BS and multiple-user equipment (UE), each equipped with a SIC receiver. An assumption is made that UE_1 is in the closest proximity to the BS, while the farthest distance location is considered for UE_K in this model. The identical signal-carrying information meant for each UE is received by both. Each UE performs decoding on the strongest signal first and subtracts it from the superposed signal. It iterates this process until it successfully isolates its own signal. However, accurate subtraction requires the original individual waveform to be accurately regenerated, which is challenging and prone to errors. For a PD-NOMA system with a K-user downlink configuration, the signal-to-noise power ratio (SNR) for the kth user, accounting for cancellation errors, can be expressed as described in reference [6] Top of Form

$$\text{SNR}_k = \frac{P_k g_k^2}{N_0 W + \sum_{i=1}^{k-1} P_i g_k^2 + e \sum_{i=k+1}^{K} P_i g_k^2}. \tag{7.2}$$

Here, the term ε denotes the remainder of the canceled message signal (cancellation error) and $N_0 W$ is the AWGN power. In the absence of cancellation errors, the third term in the denominator of eqn (7.2) becomes zero, indicating perfect SIC. The cancellation error term is changed from 1% to 10% in order to evaluate the influence of imperfect cancellation. As the cancellation error increases, both the individual user rate pairs and the overall capacity of the system decrease, as illustrated in Figure 7.1.

The influence of poor cancellation is next investigated by varying the cancellation error epsilon from 1% to 10%. As the cancellation error increases, the individual user rate pairs and the overall capacity of the system decreases as can be seen in Figure 7.1.

Machine learning or deep learning can be used to train the NOMA network so that it does not require SIC as a detection mechanism. This will remove the drawbacks of cancellation errors that are produced as a result of SIC operation, and the system throughput may increase more.

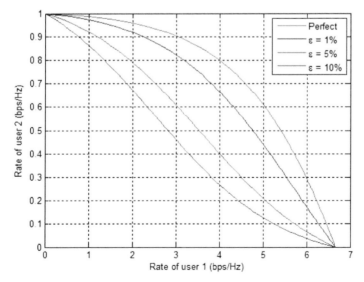

Figure 7.1 Impact of cancellation error on two-user PD-NOMA performance [6].

7.5 Interference Issues

Various techniques for interference cancellation, including SIC, are employed to handle the problem [7]. The interference cancellation is performed successively by each of the users other than the weakest channel user, to properly decode its own message. The number of SIC operations grows with the users in the network. Hence, for the multi-user scenario in PD-NOMA, such SIC is a source of a large system latency caused due to more SIC operations. The utilization of the superposition coding principle enables NOMA to accommodate multiple users sharing the same wireless resources. This approach effectively increases system capacity and spectrum efficiency by allowing the simultaneous overload of these resources by numerous users. As a result, NOMA has recently gotten a lot of press as a potential solution to some of the 5G/B5G/6G challenges. Multiple approaches exist for NOMA, one of which is the PD-NOMA. In PD-NOMA, SIC is employed for detection, although its performance enhancement potential is hindered by error propagation. A study conducted by the authors in [8] proposed an iterative interference cancellation (IIC) detection scheme called parallel interference cancellation (PIC) for uplink PD-NOMA systems. The IIC approach offers a simpler alternative to the maximum likelihood (ML)-based multi-user identification algorithm.

A revolutionary detection system based on IIC is referred to as advanced IIC (AIIC). The key distinction between AIIC and IIC is that in the first

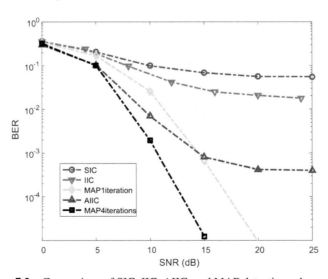

Figure 7.2   Comparison of SIC, IIC, AIIC, and MAP detection schemes [8].

iteration of the algorithm, a portion of the interferences will be removed. Following that, both algorithms' next steps are identical. On both techniques, the concerned error propagation was still present.

The simulation findings of [8] show that IIC and AIIC can outperform SIC significantly. For a fair comparison of the approaches, the maximum a posteriori (MAP) and SIC schemes were chosen as the basic schemes. A comparison of SIC, IIC, and AIIC approaches is shown in Figure 7.2.

7.6 Imperfect CSI (Channel State Information)

In the real-world scenario, obtaining the exact CSI is unfeasible, so the problem of interference and/or SIC worsens in such cases. In [9], the authors proposed a power allocation scheme that prioritizes energy efficiency in the context of NOMA with imperfect CSI. Their study focused on a multi-user downlink NOMA network and demonstrated that the energy efficiency decreases as the channel state estimation error increases. Additionally, when the number of users in the network grows, the outage probability of the system increases unless the transmission power is increased. Even with higher transmit powers, the outage probability remains significantly higher in regard to an immense number of network users. Another relevant study [10] explores the combined impact of imperfect CSI and SIC in NOMA-based satellite-terrestrial systems. This comprehensive research investigates the outage probabilities, sum rate, and throughput considering the presence of channel estimation errors and residual interference resulting from imperfect SIC. The authors highlight the strong dependence of the sum rate on the residual interference and the sensitivity of energy efficiency (EE) with residual SIC error.

To overcome the errors in channel estimation, intelligent algorithms can be employed. This will also reduce the residual errors of SIC and ensure a high EE NOMA network.

7.7 Signature Design

Inter-user interference results naturally from the multiplexing of users in the power domain using the identical resource block (RB). Conversely, CD-NOMA which utilizes sparse user signatures has a design issue with signatures. Sparse code multiple access (SCMA) uses its code word signature (from its codebook), to spread the data bits in a non-orthogonal fashion over a resource element (RE). The code words of the codebook must be sparse to facilitate the receiver with multi-user detection capability. Hence, for SCMA, the signature design problem can be studied with the SCMA codebook design issues. In [7], various approaches are presented in order to design the codebooks for SCMA systems. The codebook design is divided into three distinct problems: mother constellation (MC) design, transformation matrix (TF) design, and mapping matrix (MM) design. The transformation matrix is employed to create multi-dimensional (MD) constellations that are specific to each user. Then, to create a codebook, these MD constellations are enlarged for each user. In the optimization of MC, the goal is to maximize the average alphabet energy while ensuring a minimum Euclidean distance between constellation points [11]. Similarly, joint optimization of MC and MM is performed in [12]. Observing the fact that the system complexity may be reduced by reducing the collisions of constellation points over each dimension, low-projection MCs are proposed in a few works [13–15]. An excellent survey on the known MCs for uplink SCMA is provided in [16]. The study demonstrates that the system performance is greatly influenced by the multi-user detection (MUD) behavior across different SNR ranges. In [16], Table V presents notable 4-point and 16-point constellations for various channel conditions. However, the performance evaluations are limited to a regular SCMA system where users occupy a fixed number of REs. The SCMA designs for irregular SCMA systems have not been studied, thus, this opens an exciting research challenge for designing SCMA codebooks for irregular systems.

The signature design problem requires efficient algorithms for generating MC, MM, and TF matrices jointly. This will optimize the constellation points and offer lower error rates in an SCMA network, along with a large spreading gain.

7.8 Physical Layer Security Concerns

The physical layer security in the NOMA network becomes a significant concern as stronger channel users decode the signals of weaker users first in the SIC process. This happens when the BS pairs a trusted user with an untrusted user. The physical layer security is a complementary approach to improve network security, apart from the cryptographic technique [17]. The implementation of PD-NOMA with physical layer security aims to achieve secure communication by addressing the issue of eavesdropping. In addition to the pair outage probability (OP), which represents the reliability metric and ensures the Quality of Service (QoS) in the network, another important metric called secrecy outage probability (SOP) is considered. The SOP measures the probability that the trusted user maintains a non-negative secrecy capacity, which is determined by the difference between the capacity of the trusted user and that of the eavesdropper.

The authors in [18] analyzed this problem for a two-user PD-NOMA network. The authors showed that the pair OP increases with increasing distance of the untrusted user from the BS. For a specific trusted user distance (from the BS), the SOP is also reduced with increasing distance of the untrusted user from the BS. This SOP reduction becomes more prominent with high transmit SNR and a higher secrecy rate. Based on their findings, the authors also determined the monotonically decreasing nature of SOP for the power allocation factor. Further, the range of power allocation factors is also determined for optimal OP probability under the SOP constraint.

The eavesdropping issues are further considered in [19]. The authors explicitly mention the internal and external eavesdropping scenarios in the NOMA network. Several counter techniques for both internal and external eavesdropping issues are described that mainly rely on the beam forming techniques, apart from the appropriate power allocation technique. The eavesdropping problem in a PD-NOMA network having two-user is further analyzed in [20] from the viewpoint of secrecy fairness maximization along with the network QoS requirements. An interesting observation is a trade-off between the SOP and the OP. The authors observe that an increase in pair OP is associated with a decrease in SOP, indicating that higher rate requirements result in poorer secrecy performance. The authors determine the optimum global power allocation factor and the performance gains achieved with the optimization.

Using a relay transmission as in cooperative NOMA increases the effective available SNR at the receiver, and may ensure a significant reduction in the SOP. Hence wider coverage may be made available with cooperative NOMA, ensuring a higher secrecy rate.

7.9 Power Allocation and User-Pairing

Power allocation strategy and user-grouping issues are other issues of concern for the PD-NOMA network. Since the obtained signal power at the receiver together with the employed decoding mechanism are the two factors that primarily affect network performance, it is prime important to utilize an optimal power allocation algorithm and user-pairing algorithm so that the received signal yields an error-free message with a high probability.

Considering the various system performance metrics, viz. spectral efficiency, energy efficiency, etc., along with the analysis of the channel characteristics, and the outage, various optimization results are obtained for addressing power allocation as well as user-pairing challenges.

The joint problem of imperfect SIC and user-pairing is studied in [21]. An adaptive user-pairing approach for a downlink PD-NOMA of a two-user network has been discussed that aims at maximizing their achievable sum rate. If the SIC is imperfect, the performance of conventional PD-NOMA deteriorates. Hence the authors derive an adaptive method in regard to user pairing along with power allocation considering the residual SIC errors. Further, it is shown that an OMA-based adaptive sum rate of the network is always larger than the NOMA-based adaptive sum rate, under imperfect SIC constraints. According to the findings, which were attained using both the traditional near-far (NF) pairing approach as well as the uniform channel gain difference (UCGD) pairing method, the achievable sum rate rises almost linearly as the number of users grows. Further, the superiority of adaptive user grouping is analyzed over both the NF and UCGD pairing algorithms. However, there may exist other grouping algorithms, especially in multi-user scenarios, that are not considered. The performance of adaptive user grouping over such scenarios needs to be analyzed further.

In [22], a comparative study was conducted in two-user NOMA networks to assess various power allocation schemes. The research investigates the impact of power allocation strategies on different model parameters, including channel gain, SNR, distance from the transmitter, fairness index, and EE. Fixed power allocation (FPA), fractional transmit power allocation (FTPA), and water-filling–based allocation are a few examples of conventional power allocation techniques that are frequently employed in NOMA. It also examines a number of theoretically suggested schemes that have been discussed in the literature, including generalized power allocation (GPA), adaptive power allocation, optimum power allocation (OPA), full search, fuzzy logic, improved fractional transmit power allocation (I-FTPA), particle swarm optimization (PSO) algorithm-based allocation, and dynamic power allocation based on reinforcement learning (RL), among others [22].

Furthermore, the authors propose a novel eigenvalue-based power allocation scheme for a two-user PD-NOMA network, claiming its superiority. However, the evaluation is limited to the error-rate performance of the network at low SNRs. The study does not address the effects of the proposed power allocation scheme on other significant metrics such as the sum rate, outage performance, and fairness. Additionally, the authors focus solely on a two-user scenario where interference is manageable. In a multi-user environment, inappropriate power allocation can result in substantial inter-user interference, which can severely impact the overall system performance.

To provide massive connectivity, it is desirable to focus on multi-user NOMA systems. There is a need to re-examine the joint optimization of user pairing, user clustering, and power allocation algorithms.

7.10 High Receiver Complexity

For SCMA, the prominent decoding process is the message-passing algorithm (MPA) or its variations. The MPA involves the passing of messages between variable nodes (representing users) and function nodes (representing resources). Subsequently, the log-likelihood rate (LLR) is computed for each user based on these messages. Hence, the MPA is much more complex than the SIC decoder. Thus, to facilitate a simple receiver, lower complexity detection algorithms need to be investigated for SCMA.

Further, in the case of imperfect SIC, the residual interference has an impact on the decoding of messages from stronger channel users. This needs to be looked at for a multi-user case, as the message decoding error becomes more prominent with such error propagation caused due to imperfect SIC in a PD-NOMA system. The techniques such as IIC and AIIC use PIC detectors instead of SIC detectors, hence the receiver complexities have not been evaluated in [8]. Specifically, the receiver complexity of the AIIC technique needs to be analyzed thoroughly for a decent comparison of the SCMA receiver performance. Another work [23] discusses the error performances of minimum mean square error (MMSE) and MAP-based detectors but suggests the requirement of advanced detectors for multi-user NOMA networks. To our knowledge, the computational complexity of such detectors has not been analyzed yet for the NOMA networks.

Performance of the detector can be improved by employing techniques that utilize beamforming and combining in combination with MMSE or MAP-based detectors in a multi-user NOMA network.

7.11 NOMA Integration with 6G Heterogeneous Networks

Integration of an improved version of NOMA may be the key technology to fulfill the aspects of 6G partially or completely. To improve spectral efficiency and system capacity, it is intuitive to integrate NOMA with state-of-the-art or emerging signal processing techniques. One potential solution to address this issue is the utilization of a relay with full-duplex (FD) capability, enabling simultaneous receiving and transmission in the identical frequency range [24]. Based on this idea, several combinations of NOMA exist in literature. Some of them may be categorized as co-operative NOMA connected with FD-NOMA, cognitive radio (CR)-NOMA, and multiple-input multiple-output (MIMO)-NOMA [25].

Furthermore, the integration of cognitive radio (CR) with cooperative NOMA has garnered considerable interest because of the capability to provide high throughput, low latency, and support for a large number of connections. As both CR and NOMA rely on managing interference, their implementation must focus on minimizing interference while maximizing the utilization of available spectrum resources. This technique has shown notable improvements with regard to outage probability followed by system throughput in comparison to traditional orthogonal multiple access (OMA) systems, showcasing its potential for enhanced performance [26].

MIMO adds more spatial degrees of freedom to cooperative NOMA, allowing for system enhancement. Appropriate beamforming vectors must be constructed for the users to successfully use the MIMO gain. When the users are clustered commonly, each cluster having a near and far user, MIMO-NOMA can be utilized to boost user connectivity far more. Users in the same cluster share the same beam, whereas users in separate clusters are distinguished via MIMO beamforming [25]. A recent paper [27] shows the concept of massive MIMO cooperative cognitive relaying with NOMA, as shown in Figure 7.3.

Figure 7.3 illustrates a cluster of small cells situated within the coverage range of a macrocell BS that holds a spectrum band as the incumbent. The incumbent BS caters to multiple primary users (PUs). Each small cell, referred to as a secondary access point (SAP), is equipped with massive MIMO and NOMA interference-cancellation capabilities. This enables the SAPs to opportunistically access sub-channels within the licensed spectrum of the incumbent PUs to serve their own secondary users (SUs) dynamically. Here, the subscripts r and t in SU respectively represent the receiver and transmitter.

Hence, the idea proposed in [27] represents NOMA-based network integrated with a cognitive, cooperative relayed transmission that employs massive MIMO. Such a network might offer a significant increase in connectivity which is the fundamental requirement of upcoming 6G networks.

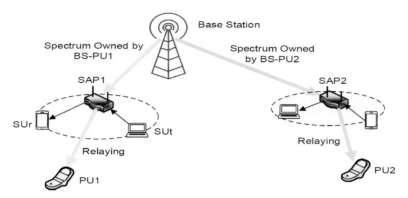

Figure 7.3 NOMA integration with cognitive cooperative relaying and massive-MIMO [27].

7.12 Conclusion

This chapter presents a basic review of the NOMA technique and summarizes the key issues in NOMA toward implementation in 6G networks. Various issues such as imperfect SIC, user interference, and imperfect CSI information have been discussed, which result in performance degradation with regard to the error rate and throughput. Viable solutions based on PIC are then discussed. Further, the signature design for NOMA, in particular, MC, MM, and TF are discussed. A proper signature design offers an additional spreading gain in SCMA systems. The secrecy issues that need to be addressed are highlighted, which are necessary to counter the probable eavesdropping problem in future networks. This is followed by a discussion on the power allocation and user-grouping strategies currently reported in NOMA literature. The optimization strategies highly depend on the performance parameter being considered, such as throughput, fairness, and energy efficiency. The algorithms for detection, demodulation, interference cancellation, and enhanced user performance require complex receiver structures; therefore, low-complexity detectors become an important concern. Finally, various cutting-edge techniques such as FD, CR, and MIMO have been reported in the chapter, which is very important to be integrated to realize 6G requirements.

References

[1] A. Holst, (2021, October), Number of IoT connected devices worldwide 2019-2030, Available Online: https://www.statista.com/statistics/1183 457/iot-connected-devices-worldwide/

[2] H. Holma, A. Toskala, J. Reunanen J, *LTE small cell optimization: 3GPP evolution to Release 13*, 2016, Wiley, New Jersey

[3] A. Mourad, R. Yang, P. H. Lehne, A. De La Oliva A, "A baseline roadmap for advanced wireless research beyond 5G," *Electronics*, vol. 9, no.2, p. 351, 2020.

[4] H. Viswanathan, P.E. Mogensen, "Communications in the 6G era," *IEEE Access*, vol. 8, pp. 57063–57074, 2020.

[5] Y. Saito, Y. Kishiyama, A. Benjebbour, T. Nakamura, A. Li, K. Higuchi, Non-orthogonal multiple access (NOMA) for cellular future radio access, in Proceedings of the 77th *IEEE Vehicular Technology Conference (VTC Spring)*, IEEE 2013, pp. 1–5.

[6] R. C. Kizilirmak, "Non-Orthogonal Multiple Access (NOMA) for 5G Networks, Towards 5G Wireless Networks - A Physical Layer Perspective, Hossein Khaleghi Bizaki, IntechOpen, vol. 10, p. 66048, 2016. Doi: 10.5772/66048. Online: https://www.intechopen.com/chapters/52822.

[7] M. Rebhi, K. Hassan, K. Raoof, and P. Charge, "Sparse Code Multiple Access: Potentials and Challenges" *IEEE Open Journal of the Communications Society*, vol. 2, pp. 1205-1238, 2021, doi:10.1109/OJCOMS.2021.3081166.

[8] M. Chen, and A. Burr, "Multiuser detection for uplink non-orthogonal multiple access system," *IET Communications*, vol. 13, no. 19, pp.3222-3228, 2019.

[9] R. Aldebes, K. Dimyati, and E. Hanafi, "Energy-efficient power allocation for imperfect CSI Downlink NOMA system," *ASEAN Engineering Journal*, vol. 11, no.2, pp. 1205-1238, 2021.

[10] J. Zhao, X. Yue, S. Kang, and W. Tang, "Joint Effects of Imperfect CSI and SIC on NOMA Based satellite-terrestrial Systems," *IEEE Access*, vol. 9, pp. 12545-12554, doi:10.1109/ACCESS.2021.3051306.

[11] M. Beko and R. Dinis, "Designing good multi-dimensional constellations," *IEEE Wireless Commun. Lett.*, vol. 1, no. 3, pp. 221–224, Jun. 2012.

[12] J. Peng, W. Chen, B. Bai, X. Guo, and C. Sun, "Joint optimization of constellation with mapping matrix for SCMA codebook design," *IEEE Signal Process. Lett.*, vol. 24, no. 3, pp. 264–268, Mar. 2017.

[13] M. Taherzadeh, H. Nikopour, A. Bayesteh, and H. Baligh, "SCMA codebook design," in Proc. *IEEE 80th Veh. Technol. Conf. (VTCFall)*, 2014, pp. 1–5.

[14] A. Bayesteh, H. Nikopour, M. Taherzadeh, H. Baligh, and J. Ma, "Low complexity techniques for SCMA detection," in Proc. *IEEE Globecom Workshops (GC Wkshps)*, IEEE 2015, pp. 1–5.

[15] F. Wei and W. Chen, "Low complexity iterative receiver design for sparse code multiple access," *IEEE Trans. Commun.*, vol. 65, no. 2, pp. 621–634, 2016.

[16] M. Vameghestahbanati, I. D. Marsland, R. H. Gohary and H. Yanikomeroglu, "Multidimensional Constellations for Uplink SCMA Systems— A Comparative Study," *IEEE Communications Surveys & Tutorials,* vol. 21, no. 3, pp. 2169-2194, 2019, doi:10.1109/COMST.20 19.2910569.

[17] A. D. Wyner, "The wire-tap channel," *The Bell System Technical Journal*, vol. 54, no. 8, pp. 1355–1387, 1975.

[18] B. M. ElHalawany and K. Wu, "Physical-Layer Security of NOMA Systems Under Untrusted Users," *2018 IEEE Global Communications Conference (GLOBECOM)*, IEEE 2018, pp. 1-6, doi:10.1109/GLOC OM.2018.8647889

[19] D. Mihaylova, V. Stoynov, A. Ivanov and Z. Valkova-Jarvis, "An overview of methods to combat eavesdropping in NOMA based networks through physical layer security," IOP Conference Series: Materials Science and Engineering, Volume 1032, no. 1, International Scientific Conference of Communications, Information, Electronic, and Energy Systems (CIEES 2020) 26th-29th November 2020, Borovets, Bulgaria, pp. 1-7, doi:10.1088/1757-899X/1032/1/012011.

[20] S. Thapar, D. Mishra, and R. Saini, "Novel Outage-Aware NOMA Protocol for Secrecy Fairness Maximization Among Untrusted Users," in *IEEE Transactions on Vehicular Technology*, vol. 69, no. 11, pp. 13259-13272, Nov. 2020, doi:10.1109/TVT.2020.3022560.

[21] N. S. Mouni, A. Kumar and P. K. Upadhyay, "Adaptive User Pairing for NOMA Systems with Imperfect SIC," in *IEEE Wireless Communications Letters*, vol. 10, no. 7, pp. 1547-1551, July 2021, doi:10.1109/LW C.2021.3074036.

[22] T. Sanjana and M. N. Suma, "Investigation of power allocation schemes in NOMA," in *International Journal of Electronics*, 2021. doi:10.1080/ 00207217.2021.1939434.

[23] C. Yan, W. Liu, and H. Yuan, "Numerous Factors Affecting Performance of NOMA for Massive Machine Type Communications in B5G Systems," *Front. Comms. Net.*, vol.2, p. 21. doi:10.3389/frcmn.2021.689530.

[24] X. Yue, Y. Liu, S. Kang, A. Nallanathan, and Z. Ding, "Exploiting full/half-duplex user relaying in NOMA systems," *IEEE Trans. Commun.*, vol. 66, no. 2, pp. 560-575, 2017.

[25] M. Zeng, W. Hao, O. A. Dobre and Z. Ding, "Cooperative NOMA: State of the Art, Key Techniques and Open Challenges," *IEEE Network*, vol. 34, no.5, pp. 205-211, 2020. doi:10.1109/MNET.011.1900601.

[26] L. Lv, J. Chen, Q. Ni, Z. Ding, and H. Jiang, "Cognitive non-orthogonal multiple access with cooperative relaying: A new wireless frontier for 5G spectrum sharing," *IEEE Commun. Mag.*, vol. 56, no. 4, pp. 188-195, 2018.

[27] Dinh, Son, Hang Liu, and Feng Ouyang. "Massive MIMO Cognitive Cooperative Relaying," in *International Conference on Wireless Algorithms, Systems, and Applications*, pp. 98-110. Springer, Cham, 2019.

Biographies

Saurabh Srivastava is currently a research scholar at the Department of Electronics and Communication Engineering, Birla Institute of Technology, Mesra. He has over 12 years of teaching and more than 4 years of research experience. He completed his graduation with distinction from the University Institute of Engineering and Technology, CSJM University, Kanpur, India, and his Masters from the National Institute of Technology, Haryana, India. He recently submitted his Ph.D. thesis on next-generation enhanced NOMA networks. Some of his initial works are published with River Publishers, Denmark. He has contributed to seven publications toward his Ph.D. in the same area, some of which are renowned SCIE and Scopus journals, and international conferences. Apart from the next-generation networks, his fields of research interest are artificial intelligence/machine learning and data science.

Rampravesh Kumar is a research scholar under the UGC-NET JRF scheme at B.I.T, Mesra, India, pursuing Ph.D. in wireless communication from the Department of Electronics and Communication Engineering since the spring of 2017. He received his B.E in ECE from RGPV, Bhopal in 2010 and M.Tech. in ECE from CMJU, Shillong in 2012. After that, he worked as NSS Engineer at Nokia Siemens Networks Pvt. Ltd. from 2012 to 2014. From 2015 to 2016, he worked as a guest lecturer in the Department of Electrical and Electronics Engineering, NCE Chandi, Nalanda. His Ph.D. work centers on cooperative communication and NOMA in improving the next-generation cellular system performance.

Sanjay Kumar has received his MBA from Pune University in 1994, M.Tech. in electronics and communication engineering from Guru Nanak Dev Engineering College, Ludhiana in 2000, and Ph.D. in wireless communication from Aalborg University, Denmark in 2009. He is an associate professor at the Department of Electronics and Communications Engineering at Birla Institute of Technology Mesra, Deoghar Campus, India. From 2006 to 2009, he was a guest researcher at Aalborg University, Denmark, where he worked in close association with Nokia Siemens Networks. From 2007 to 2008, he worked as a part-time lecturer in the Department of Electronic Systems, at Aalborg University, Denmark. Before joining teaching and research, he served the Indian Air Force from 1985 to 2000 in various technical capacities. He has nearly 35 years of teaching, research, and work experience in the field of wireless communication. He is an editorial board member of the international journal "Wireless Personal Communications," published by Springer.

8

Green Computing: Importance, Approaches, and Practices

Sunil Kr. Pandey, Kumar Pal Singh, Puja Dhar, Saurabh Saxena, and Kumar Vaibhav Bhatnagar

I.T.S, Mohan Nagar, Ghaziabad, U.P, India

Abstract

The development and utilization of information communication technology (ICT) have been enormous for the last few decades. ICT has become an indispensable part of almost all the sectors of businesses and daily lives on the globe. As per the study, the number of devices connected to ICT is 42.62 billion by 2022. Each device is made of hazardous material and the majority of the users are either unaware or they are least bothered by the utilization and required disposal of the devices. With the rapid increase in the number of devices, there is a significant rise in energy consumption, which becomes a cause of multiple factors of atmospheric hazards such as excess production of CO_2, excess accumulation of electronic waste, global warming, and depletion of the ozone layer at large. With the apprehension of these issues, there is a requirement to take measures for the solutions to mitigate all these issues. The solution is green computing. The objective of green computing is to apply computing resources with insignificant influence on the environment to ensure that all specialized, budgetary, and social goals are fulfilled. Green computing ensures a balanced and doable approach together with the imaginative needs of current and future times without compromising characteristic science. This chapter aims to uncover the various dimensions including optimization of energy consumption, approaches to reduce biological hazards (H/Gasses, etc.), and effective techniques to design environment-friendly frameworks with recycled uses of hardware and the applications of green computing. The chapter also emphasizes the latest techniques and approaches of green computing, green design, and developments, and approaches for energy optimization of data centers. The chapter also discusses the present industry standards of green computing and the renewal of green resources of energy. The chapter concludes with some case studies of green computing and open issues and challenges in the field of green computing.

Keywords: Green Computing, Green Energy, Energy Optimization, Virtualization, Virtual Server, Optimization

8.1 Introduction

Due to the huge deforestation, combustion of fossil fuels, and rapid industrialization, there have been significant changes in temperature and weather patterns over the past few decades. As a result, both the ocean and the air are now warmer on average. Sea levels have risen as a result of more snow melting as a result of rising air temperatures [19].

For society to have a sustainable future, the rising energy demands around the world and the increasing depletion of fossil resources have been identified as critical challenges. This insight has sparked a push for green cloud computing, which is crucial for enabling and empowering the move to a more feasible society with a smaller carbon footprint [20]. Green computing is usually called green IT or ICT maintainability. According to San Murugesan et al. [2008], green IT is the investigation and honing of productivity and successfully planning, making, applying, computer arrangement, servers and other subsystems such as monitors, printout devices, storage devices, and organizing communication frameworks.

The study in computers demonstrates that carbon dioxide (CO_2) and other pollution are harming the environment and the world's climate. Because it tries to preserve life, preserving our beautiful planet is an important and valid goal. The study in ICT demonstrates that carbon dioxide (CO2) and other pollutants are harming the environment and the world's climate. Because preserving our beautiful planet is an important and valid goal, researchers and professionals are paying close attention to reduce ewaste and usage non-toxic materials.

8.2 Evolution of Green Computing

Nowadays, massive amounts of data, including business transactions, phone records, and other information, are analyzed and recorded by IT systems. To run the servers and keep them maintained, all of this ongoing data storage in data centers and configurations like data warehouses require a significant amount of power. The amount of energy needed to create, maintain, and decrease the temperature of computation devices has increased momentously in current years, resulting in staggering numbers like the 60.5-billion-kilowatt hours of electricity consumed annually, which is estimated to cost \$4.2 billion. These configurations are unsustainable given the impending energy crisis and the volume of improper energy use.

In 1990, Joined Together States Natural Security Office (JTSNSO) started Vitality Star, an international labelling program to advance and recognize energy efficiency in screens, climate control, and other innovations. As a result, consumer electronics adopted sleep mode on a large scale. It is likely that the phrase "green computing" was first used soon after the Energy Star program started. The TCO certification program was subsequently introduced by the Swedish company TCO Development to support computer displays with low magnetic and electrical emissions. Later, this program's requirements were broadened to cover energy usage, biotechnology, and the use of harmful ingredients in buildings. This opened the door for green computing to be seriously considered on a global scale.

8.3 Green Computing Techniques and Industry 2023

After the US and China, India occupies the third position in the production of e-waste globally, and the rate of volume increase is steadily increasing. Computer equipment makes up about 70% of India's e-waste, followed by 12% in telecom, 8% in electric devices, and 7% of the devices used in the medical industry, according to a report by KPMG and ASSOCHAM. India is among the worst five nations and ranks 177 out of 180 on the 2018 Environmental Performance Index. It was associated with worse outcomes in environmental health and air pollution-related mortality. Many states in India are generating e-waste and its approximate estimation is shown in Figure 8.1.

Numerous technologies exist that are specifically created to increase people's awareness of their energy use and encourage them to alter their behavior in favor of more sustainable energy use. Some technologies aim to convince people to reduce their energy consumption, while others convince them to shift their energy consumption to off-peak or low-demand periods (Gooding Dan et al. [2006]), and still, others convince them to shift their energy consumption to periods when it is "green," that is when it is generated from renewable resources like wind and solar power.

Figure 8.1   Generation of E-waste generation in some of the states in India.

In India, many agencies conduct different research, many researchers are conducting several studies to discover the quantity of e-waste amount of manufacturing in our country. Mostly the studies are framed on prototypes of the undesirability of electronic products. In 2005, one report was produced by the Central Pollution Control Board which said that approximately 8.5 lakhs of e-waste were produced in 2012 and which is expected to increase more in the upcoming years.

8.3.1 Monitor the usage of energy consumption

There are many data centers in the country which are having huge power consumption. It is advised to monitor the overall consumption of power by dividing the usage into how many actual servers are required for computing, data storage, and networking. Policies should be followed by each center for better management of power consumption.

8.3.2 Substitution of more power-consuming devices with less than one

Many organizations are using devices that consume more power and generate more heat. Organizations switch to HVAC technologies mostly known for heating, ventilation, and air conditioning. This technology is mostly used to regulate humidity and temperature in the devices.

8.3.3 Usage of virtual servers

Industries are moving from physical servers to cloud computing. To decrease the consumption of power, industries are applying virtualization where many servers are monitored by one machine. By virtualization data centers improve their infrastructure by using less electricity. After using the virtual server industry can also enjoy the left-floor space. Virtualization also minimizes the energy consumption. Industries are also moving toward a green grid.

8.3.4 Green disposal

Industries are producing a lot of e-waste by switching from old technology to emerging technologies. Old systems are replaced with new ones. Revamping and reusing old computers and disposing the unused systems with safe techniques like incineration is applied.

8.3.5 Natural cooling system

We know that data centers are dealing with the processing of voluminous data which results in generating more heat. The survey said that cooling the servers needs 1-1.5 W. of power used (Goodin Dan et al. [2006]). It is said as the data centers increase, the cooling powers to these servers will also increase. Many data centers are opting for natural cooling systems, by setting them up in cooler locations.

8.4 Green Computing Toward Economic Development

Green development means promoting development and economic growth while making sure that natural resources continue to provide the resources and ecosystem services that are primarily needed for our well-being.

To discover a solution to balance the contradiction between economic expansion and the environment's limited resources is the goal of the green economy. This section outlines the path to achieving sustainable development and ways to continue green development by analyzing the current domestic and international conditions. Economic development is governed by ecological environment tolerance that makes sure of economic reproduction which is expanded based on natural reproduction. The growth of the economy of a nation and the environment to create new economics is based on the market as the direction and uses conventional industry economics as the foundation. The industrial economy is evolving to meet the demands of the environment and human welfare.

Cloud computing is a mix of computing principles in which millions of computers connect in the real world and real-time offering the user with a seamless experience, as if they are accessing a single massive resource. This type of system offers numerous services such as web data repositories, massive computer resources, data processing servers, and so on.

Atrey et al. [2013] find another key issue that consumes a huge amount of energy in data centers is the cooling of machines. Historically, cooling was done by using mechanical refrigerators that supply cool water to IT hardware. Nowadays, pre-cooling, which is also known as free cooling, is used. Mechanical cooling is significantly reduced while using free cooling. For example, Facebook locates its data center in Sweden, which already has a cold and dry environment. Alternatively, Microsoft keeps servers out in the open area to allow for easy cooling. Google, also uses river water to cool its data centers. There are several hardware technologies, such as virtualization, and software technologies which beautifully introduce important concepts such as virtualization, power management, material recycling, and telecommuting for green cloud computing. The main objective of this study is to reduce overall energy consumption by consolidating or scheduling tasks as well as resources used in green cloud computing. The good findings revealed in the research are not for direct extreme energy reduction, but rather for possible electricity savings in large cloud data centers. The demand for cloud computing is rapidly expanding, as is the use of energy and the emission of hazardous gases, which is extremely damaging and a major concern in the field of healthcare, as well as a major cause of the increase in the cost of operations. Three strategies are being used to revolutionize the future of green cloud computing.

8.4.1 Nano data centers

Nano data centers consume much less energy as compared to traditional data centers. Nano data centers help to reduce heat dissipation costs and have high service proximity. They are self-capable of adaptation and scalability.

8.4.2 Scaling of dynamic voltage frequency

It is a strategy that uses frequency scaling to reduce energy use and power consumption. Using this strategy will cut energy usage and maximize resource utilization.

8.4.3 Virtualization

Virtualization is a technology that enhances machine management and energy economy by allowing several end users or organizations to share a single instance of a resource and application at the same time. Virtualization increases the number of system resources for the environmental benefit.

Haripriya et al. [2022] propose that minimizing the company's total power used does not mean only powering down your computer or workstation lights when it is not used. When your firm runs on-premises servers, although this makes a significant impact, you must be aware of the weight of consumed power. We can lessen your reliance on these on-premise servers after migrating to the cloud, which means you will need little equipment in the office, which will necessitate less room and cooling, resulting in a lower rate of power use. Savings from lower capital expenditures can be used to fund further environmentally friendly projects or corporate expansion efforts, such as strengthening marketing campaigns.

The world becomes "smarter" as science and technology advance at a rapid pace. Smart IoT devices like watches, mobile phones, and laptop computers will serve every person, IoT-enabled transportation systems like automobiles, buses, trucks, and trains, and smart environments places like homes and offices in such a smart world. Let us take an example, suppose if you use a GPS, a person's location can be continuously transmitted to a GPS server, which rapidly transmits the optimal path to the trip's destination, guaranteeing that the person is not gone. You may become stuck in traffic.

Meghashree et al. [2019] found that the audio sensor on an individual's mobile can detect a speech disparity and transmit it to a server, which compares the difference between them to a set of fingerprints to figure out if the person is sick or healthy. Lastly, all components of individual cybernetic, social, physical, and spiritual worlds are taken into account and will be intelligently interconnected in the intelligent world. Universities, companies, governments, and other organizations are all focusing on the intelligent world as the next level of critical change in human history.

The green Internet's promising future will transform our surroundings into healthier and greener environments, with high-quality services that are socially, environmentally, and economically friendly. The most exciting topics today are those that relate to the environment, including communication and ecological systems, green development and implementation, environmentally conscious Internet services and applications, energy-saving technologies, RFID, and networks.

To create effective and economical greening IoT solutions, the following research disciplines had to be investigated:

(i)   UAVs must replace an enormous amount of IoT devices, particularly those used in traffic, agriculture, and surveillance, thereby lowering energy usage and overall pollution. UAVs are a potential technology that provides low-cost, high-efficiency green Internet.

(ii) Sensor data transfer is more valuable in a mobile-cloud scenario. The wireless sensor network and the mobile cloud are combined in the sensor cloud. It is a very promising eco-friendly technology. Green social network as a service (SNaaS) research looks into the aspects of system energy efficiency.

(iii) M2M communication is critical for lowering energy consumption and hazardous pollutants. To activate automated systems, smart devices must get smarter. In the case of traffic, the machine's automatic time shall be maintained at the lowest value, and the necessary and immediate remedies should be taken.

(iv) Sustainable Internet of Things patterns can also be carried out from viewpoints that give excellent performance and service quality. Looking for the right approaches to increase service quality characteristics (such as time, bandwidth, and performance) can boost eco-friendly IoT effectively and efficiently.

(v)   As IoT grows, it consumes very little energy, lowers the total negative impact of IoT on people's health, seeks new sources, and alters environmental conditions. Green IoT can then make a substantial contribution to a more sustainable and smart environment.

(vi) To achieve energy equilibrium, radio frequency energy collection should be considered while communicating with IoT devices.

(vii) More research is required to develop IoT devices that help decrease CO_2 emissions and energy usage. The fundamental goals of a smart and sustainable environment are to save energy and reduce CO_2 emissions.

8.4.4 Storage and management technology for big data

Memory storage and the development of the relevant database are required for the management and transfer of huge data. As a result, emphasis has been placed on technology management and processing for complicated structured, semi-structured, and unstructured big data. As a result, it is critical to address challenges such as massive data display, dependability, storage, transmission, and affective processing. Memory storage and the development of the relevant database are required for the management and transfer of huge data. As a result, emphasis has been made on technology, managing and processing complicated, structured, semi-structured, and unstructured big data. So it is critical to address challenges such as massive data display, dependability, storage, transmission, and affective processing.

Non-relational databases, database caching systems, and relational databases are the three forms of database technologies. Relational databases are classified into two types: traditional relational database systems and NoSQL databases. Meanwhile, non-relational databases are known as NoSQL databases and are classified as image storage databases, key databases, document databases, and column databases. Furthermore, the advancement in security for big data technologies is required. Transparent encryption, data auditing, data destruction technology, and data access control are all useful tools in this regard. Furthermore, privacy protection, data verification, inferential management, forensics, and data authenticity detection, were all improved.

8.4.5 Data analysis and mining technology for energy consumption

Big data from the manufacturing sector receives information based on the level of energy emission, which may be utilized to extract potentially important data and knowledge. This data type is fuzzy, noisy, implicit, random, and incomplete. Data mining is classified according to its applications as predictive models, data summaries, association rule discovery, exception identification, clustering, and sequential pattern discovery. Furthermore, they are classed as relational databases, text data, heritage databases, spatial databases, multimedia databases, and temporal databases based on the objects of mining. However, they are categorized as statistical methods, machine learning approaches, neural networks, and database methods by mining methods. Statistical processes include regression analyses, discriminant analyses, exploratory analyses, and cluster analysis. Neural networks include forward neural networks and self-organizing neural networks. Inductive learning techniques, genetic algorithm models, and case study approaches are examples of machine learning approaches.

Making artificial intelligence green as well as sustainable or the AI green approach requires a bias-free (other than a positive discrimination or reasonable environmental bias), inclusive, explainable, trustworthy, ethical, and responsible approach to technology that aims to alleviate the planet's developmental challenges sustainably. This type of approach which uses AI to solve sustainability challenges more sustainably – will also catalyze smart city development.

Yigitcanlar et al. [2021] proposed to reduce sensor energy consumption, under the umbrella of WSNs, a range of approaches and technologies are created.

These ideas have naturally been applied to the IoT paradigm. Wireless sensor networks and IoT have been used in a range of applications, like detection of forest fires, environmental monitoring and protection, and ambient monitoring. The fundamental reason for inventing these strategies was to decrease the amount of energy needed primarily for sensing devices, that are mostly wirelessly interconnected and battery operated.

Fog and edge computing-based solutions are most likely the most significant advancements in this area, which decrease data transfer along with bandwidth requirements of IoT-based applications by calculating data at the fog or edge layers, which are close to sensors and devices. This also provides other benefits like privacy and data security. Many studies have been carried out to assess the energy savings and other advantages of edge and fog technology.

8.5 Energy Optimization at Data Centers

Data centers are places where large numbers of servers are placed to cater requirements of various computing services. In the present context of diverse information communications and technology (ICT) requirements, data centers have emerged as the primary load bearers. Along with source of services solutions, data centers now are becoming a point of concern for high-level energy consumption which further leads to several environmental hazards. Now some of the crucial consequences of high-level energy consumption by data centers are significant production of CO_2, excess heat generation, global warming, and depletion of the ozone layer. According to statistics, data centers used 140 billion kWh of energy globally in 2005, which was about 1% of the total amount used globally. By 2017, the consumption of energy increased by 416 TWh worldwide which is 3% of worldwide power consumption, which is expected to rise by more than 2000 TWh by 2030. In the present scenario of the digital world, data centers currently use around 3% of the world's energy, but by 2030, that percentage is predicted to increase to 8%. Therefore, the concerned organizations have already started examining their data centers' energy footprint to find methods to optimize energy consumption Rong et al. [2016] and Murugesan et al. [2008].

There are multiple factors for power consumption at data centers such as computation, communication equipment, storage devices, other physical resources, and cooling systems. The division of power consumption in various parts of a data center is presented in Figures 8.2 and 8.3.

The two biggest energy-utilization factors in data centers are heat emission control systems and computing. Optimization of energy for these two factors becomes more significant to decrease the energy utilization at data centers.

8.5.1 Technologies and approaches used for energy optimization at data centers

The approaches used to optimize the power consumption at data centers fall into five categories: (i) reducing energy consumption by IT infrastructure including computer equipment, servers, and storage; (ii) power infrastructure reducing power wastage at power sources and distributed equipment; (iii) airflow management by improving cooling by utilizing natural airflow mechanisms; (iv) HVAC: optimizing cooling and humidification systems; and (v) others: it covers other factors responsible for energy consumption. Some of the efficient and popular technologies and approaches are listed below (Katal et al. [2022]):

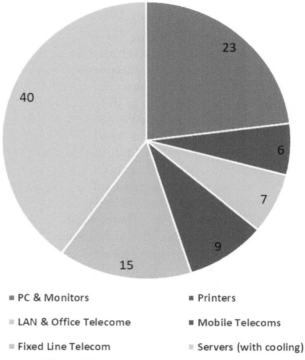

PC & Monitors   ▪ Printers

LAN & Office Telecome   ▪ Mobile Telecoms

Fixed Line Telecom   Servers (with cooling)

Figure 8.2   Estimated ICT CO_2 emissions.

8.5.2 Energy optimization for IT infrastructure

There are many ways to optimize the energy consumption in IT infrastructure at data centers including:

i) **Consolidate lightly utilized servers** Usually, the server utilization at the data center is not optimized. Most of the time the server remains underutilized. As per the study, severe with no load or very low load consumes 70% of the energy of its full load. Consolidation of servers provides energy optimization and load balancing on the servers.

ii) **Virtualization**
    A virtual server is an emulation of a real server that is available online through the web which works and behaves like a real server. The virtualization technique of the server provides a platform for servers' consolidation. It allows running of parallel different workloads on one physical or host server. The virtualization approach optimizes energy

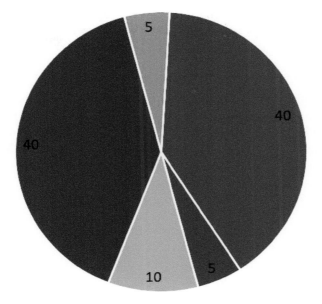

- ■ Server power consumption
- ■ Communications equipment energy consumption
- ≡ consumption Energy consumption of power supply system
- ■ Energy consumption of refrigeration system
- ≡ Energy consumption of the storage device

Figure 8.3 Distributions of power consumption at data centers.

consumption by reducing the number of physical servers at the data center.

iii) **Clustering servers**

The server clustering technique reduces the number of redundant servers. Usually, standby servers are installed at data to handle failover situations. If the number of servers in the cluster including stand-by servers is not optimized, it can lead to more energy consumption.

iv) **Downsizing the application portfolio**

It has been observed in data centers that duplication of applications is installed and multiple versions of very similar applications are installed on servers without removing their previous versions. The duplication of such application also consumes significant resources of the server and cause unnecessary power consumption of the servers. A smaller number of servers is required when redundant applications are removed.

8.5.3 Implement efficient data storage techniques

The amount of data stored in data centers is enormous. There is a requirement for smart and dynamic strategies to store and access the data so that its storage and access need less energy consumption. The strategies for handling data in optimized ways include the use of existing hardware, and reducing the size of data by compressing it or by removing redundant data. Some important data storage strategies are briefed below:

i) Storage tiers

Tiered base storage of data using different layers of storage media can help to reduce the total storage cost. The number of tiers may depend on a trade-off between the price of the storage medium, the required performance, and the frequency of data access. The best-performing storage is chosen to store the data that is used most frequently. Data that is not used as often is kept on less expensive or low-performance storage.

ii) Virtualizing storage

Data storage using storage virtualization is done in a distributed method. The data can be stored at various locations on different servers which look like storage at a single physical server. Storage virtualization improves storage performance, enables storage tiers to enable, and makes it simpler to increase storage capacity.

iii) Software-defined storage (SDS)

Important aspects of software-defined storage environments like thin provisioning, de-duplication, and snapshots are featured by storage virtualization. Large data storage utilization reduces the overall amount of storage equipment needed for data centers, which lowers the operational and cooling energy requirements.

8.5.4 Utilize built-in server power management features

Usually, the newer servers are more energy-efficient than older servers. These features generally include better DC voltage regulators, more efficient power supplies, less power-consuming processors, energy-efficient cooling systems, and built-in energy management mechanisms.

8.5.5 Control airflow for effective cooling

Managing airflow is a straightforward task. The flows of hot as well cold air are to be kept separate. The hot air flow should not affect the flow of cold air and vice versa. The flow of cold air should not be mixed with the flow of hot air, hence inlets of cold air should be kept separated from the exhaust of hot air (Hang et al. [2021]).

At present the data centers contain a higher number of server racks than past time. As a result, the single server rack consumes up to 60 kW in comparison to the earlier case of per rack power consumption which was near to 1–5 kW. The heat emission ratio has increased up to 10 or more times where airflow management becomes more important.

8.6 Industry Standard for Green Computing

Standards are benchmarks in any of the fields to maintain adequate quality of the product and services. Some important standards in modern information technology guide and maintain the development and sustainability of green computing in various dimensions. Some of the important standards for green computing are EPA Energy Star, EPEAT, RoHS, WEEE, and SpecPower (Murugesan et al. [2008]).

8.6.1 EPA Energy Star

The Energy Star standard was created by the Environmental Protection Agency (EPA). This is a standard for energy efficiency which is run by the United States government in collaboration with EPA and the Department of Energy. The concerned organization also run some certification program for this standard which covers a wide variety of electrical devices including home appliances, office equipment, heating and cooling systems, and lighting. Energy Star specifications for IT devices were designed to manage power so that CPUs, disc drives, and display units may go into a low-power mode while not in use without losing functionality (Harmon et al. [2009]).

8.6.2 RoHS

The Regulation of Hazardous Substances (RoHS) of the European Union is in charge of limiting the use of hazardous components in electronic items. The manifestation of several hazardous substances like hexavalent chromium, mercury, lead, cadmium, polybrominated biphenyls, and polybrominated diphenylethersis are tested for in agreement with RoHS. The RoHS standard guarantees that any new equipment you buy and use should be free of the hazardous materials indicated above.

8.6.3 WE

The proper treatment of electronic trash (sometimes known as "e-waste") is governed by the European Community regulation 2002/96/EC, or the Waste Electrical and Electronic Equipment (WEEE) directive. WEEE ensures the safe disposal and safe recycling of such e-waste by the original producer of the electronic equipment, by taking required measures by them.

8.6.4 SpecPower

Software benchmarks developed by the Standard Performance Evaluation Corporation (SPEC) are used to assess the performance of computing systems. SPEC also has a benchmark known as SpecPower which evaluates the power versus performance characteristics of computing servers (Saha et al. [2014]).

8.7 Renewable Green Energy

The energy that is generated from energy production sources that are naturally renewable like sunlight, water, and wind flow is called green energy. This type of energy is produced without harming the environment and energy-generating sources are almost never-ending. One of the prime advantages of green energy is that its sources are harmless to the environment and it does not generate greenhouse gases (Kumar et al. [2010]).

8.7.1 Working of green renewable energy

Renewable energy sources like solar, wind, geothermal, biomass, and hydro-electricity are commonly used as a source of green energy. Whether it is through solar panels, wind turbines, or hydraulics, each of these technologies produces energy differently. A resource must not emit pollution to be deemed green energy, such as fossil fuels. This means that not all forms of renewable energy are favorable to the environment. Because of the CO_2 released during the burning process, power generation that uses organic material from sustainable forests, for instance, may be renewable but is not necessarily environmentally friendly. In contrast to fossil fuel sources like natural gas or coal, which can take millions of years to develop, green energy sources are often regenerated organically. Mining and drilling operations, which can be detrimental to ecosystems, are also commonly avoided by green sources (Kalyani et al. [2015]).

8.7.2 Types of green renewable energy

The main sources of energy are wind, solar, and hydroelectricity including tidal energy, which harnesses ocean energy from tides in the sea. Electricity from the sun and wind can be produced on a small scale in individual homes or on a larger, industrial scale (Kalyani et al. [2015], Majid et al. [2020]). The six most common forms of green renewal energy are listed as follows:

i) Solar power

Photovoltaic cells, which capture sunlight and transform it into electricity, are frequently used to produce common types of renewable energy. Additionally, solar energy is used to cook, light, and heat buildings as well as to provide hot water. Solar energy is now affordable enough for personal use, such as garden lighting, and it is also used on a larger scale to power entire neighborhoods, as seen in Figure 8.4.

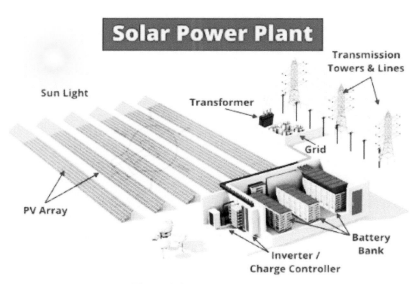

Figure 8.4   Solar power plant.

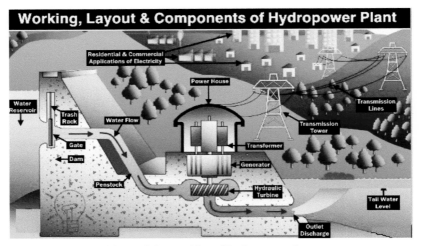

Figure 8.5   Working of hydropower plant.

ii) Wind power

Utilizing the power of the world's air movement to propel turbines that then produce electricity, wind energy is especially well-suited to offshore and higher altitude sites.

iii) Hydropower

The flow of water in rivers, streams, dams, and other bodies of water is used to generate electricity in this type of green energy, also known as hydroelectric power (Figure 8.5). On a small scale, hydropower can be produced by running water through household pipes, or it can be produced by evaporation, rainfall, or ocean tides.

iv) Geothermal energy

Utilizing thermal energy that is just below the earth's crust, this type of green energy is produced. Although this resource requires drilling to access, raising questions about its effects on the environment, it is a vast resource once it is used. The same resource that has been used for thousands of years to heat hot springs can also be used to run turbines and produce electricity. The energy stored beneath only the United States is enough to produce ten times as much electricity from coal as it does right now. While some nations, like Iceland, have easy access to geothermal resources, drilling techniques must always be controlled to be completely green because the ease of exploitation of these resources depends on geography.

v) Biomass

For renewable resources to be accepted as a credible "green energy" source, it must also be treated properly. Burning wood waste, sawdust, and combustible organic agricultural waste is how biomass power plants produce energy. While these materials do emit greenhouse gases when they are burned, they do so at much lower levels than fuels based on petroleum.

vi) Biofuels

The organic resources can be transformed into fuels like ethanol and biodiesel, as opposed to being burned as biomass. In contrast to only providing 2.7% of the world's transport fuel in 2010, biofuels are anticipated to fulfill more than 25% of that demand by the year 2050.

8.7.3 Importance of green renewable energy

Green energy is essential for the environment because it replaces fossil fuels' detrimental effects with more environmentally friendly alternatives. Green energy produced from natural resources is typically clean, and renewable, and generates no or very little greenhouse gases. It is also frequently accessible.

Even when a green energy source's complete life cycle is taken into account, it generates much fewer greenhouse gases than fossil fuels and little to no air pollution. This is not only great for the environment, but it's also better for the health of everyone who has to breathe the air, including people and animals.

Due to its local production and resistance to geopolitical crises, price increases, and supply chain disruptions, green energy can also result in stable energy prices. The economic advantages also include the creation of jobs in the building of structures that typically benefit the areas where the workers are employed. As a result of renewable energy 11 million jobs were created worldwide in 2018, and as we work to achieve objectives like net zero emissions, this number is projected to increase in the coming time.

8.8 Research Contribution to Green Computing

Green computing focuses on designing, developing, and promoting environmentally sustainable computing systems and practices. Earlier researchers have worked on energy-efficient computing, renewable energy sources, green data centers, e-waste management, green computing policies, and many more in the area of green computing. In the fields of the Internet of Things (IoT) and smart cities, cloud computing, artificial intelligence (AI), machine learning, and blockchain technology, researchers and practitioners are applying green computing. In this section, we will mention the major research contributions from 2015 onwards.

Because of the increasing cost of ownership for computing infrastructures in many businesses, cloud computing providers that offer on-demand infrastructures have become more popular. Ankita et al. proposed a method in 2015 for optimizing virtual machine placement by live migration applying dynamic threshold values to provide a deadlock-free resource allocation that focuses on multidimensional resources to decrease the energy consumption of data centers. Zoha et al. (2015) provide a complete review of the most recent VM placement and consolidation approaches to improve the energy efficiency of green clouds. In his work, Debnath et al. (2015) described how to manage e-waste, or the hardware component of computers, and viewed it as a metric for green computing.

By equipping a new generation of professionals with the abilities and knowledge necessary to address the environmental challenges associated with ICT, according to research by Alexandra et al. [2016], the establishment of an international master's program in green ICT will significantly advance sustainable development. Cloud computing can potentially reduce the environmental impact of ICT, but its overall impact depends on several factors, including the energy sources used to power data centers, the carbon intensity of those energy sources, and the efficiency of data centers in terms of energy use. André et al. [2016] proposed a framework for environmental accounting of cloud computing, which takes into account the direct and indirect environmental impacts of cloud computing. They assert that their methodology might be used to assess the environmental impact of cloud computing and compare the environmental performance of different cloud service providers. Mobile cloud computing may be able to reduce the energy consumption and carbon footprint of mobile devices by moving processing and data storage to cloud servers. The model put forth by Keke et al. in 2016 makes use of cloudlets, which are tiny cloud data centers placed close to mobile devices. The workload and energy availability allow the cloudlets to dynamically modify their energy consumption.

Huigui et al. [2016] reviewed existing methods for energy optimization in data centers, including server consolidation, workload balancing, and power management. The authors then put up a fresh strategy for reducing energy use in data centers that involves dynamically modifying the supply voltage of servers based on workload parameters.

The industrial Internet of Things (IIoT) requires more expensive data centers for green computing, hence Guangjie et al. [2017] proposed a resource-utilization-aware energy-efficient server consolidation algorithm. To choose the best server consolidation technique, the suggested algorithm takes advantage of resource utilization measures, such as CPU utilization, memory utilization, and network bandwidth utilization. A fuzzy logic-based solution is also suggested by the authors to deal with workload variability and uncertainty in IoT contexts.

The current state of research on server consolidation and virtual machine migration strategies in DVFS-enabled cloud data centers is thoroughly analyzed by Mirsaeid et al. [2018]. Irfan et al. [2018] developed an edge/cloud computing strategy for workload-aware virtual machine (VM) consolidation in the context of Internet of Things (IoT) applications. The author proposes a heuristic algorithm that can efficiently handle the dynamic and heterogeneous workloads in IoT applications. Laith et al. [2018] proposed approach involves the use of a centralized controller that schedules the paths and messages between IoT devices based on their energy consumption and availability.

Adedapo et al. [2019] surveyed IT professionals in Malaysia to assess their green beliefs, attitudes, and perceptions toward green computing practices and highlighted the importance of promoting green computing practices among IT professionals and the need for organizations to provide the necessary support, training, and resources to enable IT professionals to adopt these practices. Yixiong et al. [2019] proposed an approach that involves the use of an intelligent green scheduling algorithm that optimizes the scheduling of jobs in the workshop, taking into account machine states and energy consumption. The authors also suggest a strategy based on edge computing that allows for real-time monitoring of machine statuses and energy usage, which can help lower energy consumption and enhance the workshop's overall sustainability.

A study on energy-efficient power management techniques was presented by Ankit et al. [2020]. Due to numerous advantages including automatically optimized resource management and cutting-edge service delivery models, cloud computing is being quickly accepted for managing IT services as a remarkable option. Niloofar et al. [2020] argue that traditional VM consolidation techniques may not be sufficient for energy optimization, as they do not take into account the growing trend of container-based virtualization. As a result, they propose a joint approach that consolidates both VMs and containers, taking into account the resource requirements of each workload. Sururah et al. [2020] identified obstacles to greater cloud computing use in the construction sector and discussed solutions to these obstacles.

Monika et al. [2020] used a classification methodology to analyze and categorize the existing literature on green IS adoption. The authors present that the main drivers for green IS adoption are regulatory requirements, corporate social responsibility, and cost savings. The authors also identified several strategies for promoting green IS adoption, such as employee training, green procurement policies, and the use of environmental management systems.

An overview of the concept of green ICT and its importance in addressing the environmental challenges associated with the growth of ICT infrastructure, Sakshi et al. [2020] discussed the potential of 5G-NB-IoT in enabling green ICT by supporting low-power and low-bandwidth applications. They covered the potential of 5G-NB-IoT in enabling various sustainable applications, such as smart cities, precision agriculture, and industrial automation.

A task scheduling optimization strategy for green cloud computing was put forth by Wanneng Shu et al. [2021] to enhance system performance while lowering energy use and carbon emissions. When assigning jobs to the cloud resources that are available, they proposed an algorithm that takes into account both the execution time and energy usage of each work. The program also considers the carbon emissions linked to the resource's energy use. Jaiswal et al. [2021] studied the usage of green computing in the context of the Internet of Things (IoT) using a technique called time slotted simultaneous wireless information and power transfer. The simultaneous transport of data and power across IoT devices is made possible by this method, which could result in more energy efficiency and power between IoT devices, which can lead to more energy-efficient and sustainable operations.

8.9 Case Studies of Green Computing

Problem: Data Center consumption of power

Solution and Devices: Thin clients are the type of machines that allow users to access as well as record data on a central server which can be located anywhere in the world, using only a display screen, keyboard, and mouse. It assists consumers in eliminating the requirement for multiple tasks such as anti-virus, firewall, and upgrading. Thin client advantages involve: cheap infrastructure costs, maintenance is substantially lower than in a traditional multi-computer system.

Problem: Energy usage toward data centers accounts for about half of data center running expenditures, putting significant strain on the environment.

Solution: Adoption of techniques for green computation:

 i) Virtualization is related to the generalization of computer resources. In such cases, virtualization is necessary to merge many physical systems into virtual equipment, as well as detach the original hardware and decrease power and cooling costs.
 ii) Recycling of materials to reduce system needs for desired equipment.
 iii) Energy conservation by allowing the system to turn off IT components regularly.
 iv) Consideration of a data center with green computation, as well as its associated systems and data storage, may necessitate a power supply and security system. It may feature effective system management with minimal power usage (Analyzing Green Computation Linked to Cloud Resources: A Case Study. In *International Conference on Artificial Intelligence in Manufacturing & Renewable Energy (ICAIMRE):* Prasad et al. [2019]).

Problem: Power-guzzling servers

Solution: The power used by each server, in addition to the total amount of energy used by all connected servers, is reanalyzed by the power analyzer. The amount of electricity used was monitored and recorded every ten seconds. The calculated power consumption data is averaged annually to obtain the total year of power consumed by each server and, as a result, the total year of energy consumed by all connected servers. To decrease this cost, it is recommended that a net-metered solar photovoltaic national grid hybrid system shall be used, that will function with present servers or with traditional servers being replaced by cloud servers. This will conform with the two situations listed below:

First of all, a hybrid photovoltaic solar power for the national grid system is proposed, using current servers.

Second, an integrated PV solar-national electricity grid that replaces existing servers with cloud servers.

Photovoltaic solar is strongly encouraged to be used in addition to the existing electrical source of power to provide the necessary electrical power to the center (net metering).

It is suggested that the old 169 convectional servers shall be replaced with cloud servers that have the same technical specifications which are same as the three current cloud servers. Both systems' payback times are the same at 20 months.

8.10 Conclusion

In the present era commuting services are required at a huge scale. Almost every facet of life has become dependent on computing services and the usage of data either in online or offline mode. To meet such huge computing requirements the ICT, Information Communication Technology has laid down suitable and adequate infrastructure in terms of local network setup, medium-level computing setup, distributed network system, cloud computing, IoT, big data, and large data centers worldwide. Every computing setup requires the consumption of significant energy. The excess use of energy becomes a cause of environmental hazards in multiple ways. The environmental hazards further lead to many crucial consequences like CO_2 emissions, global warming, e-waste at large scale, and ozone layer depletion at large. The concerned stakeholders now have started to think about the serious concerns of environmental issues. The outcome is the evolution of green computing which is a way to mitigate the negative sides of ICT in terms of excess energy consumption which further leads to many environmental implications. In this chapter, efforts are put to highlight some important aspects of green computing. The chapter discusses an overview of green computing, approaches and techniques used in green computing, how energy utilization can be optimized at data centers, and what are sources of renewable green energy. The chapter concludes with issues and challenges and some case studies in the field of green computing.

References

[1] Rong, H., Zhang, H., Xiao, S., Li, C., & Hu, C. (2016). Optimizing energy consumption for data centers. *Renewable and Sustainable Energy Reviews*, *58*, 674-691.

[2] Kumar, A., Kumar, K., Kaushik, N., Sharma, S., & Mishra, S. (2010). Renewable energy in India: current status and future potentials. *Renewable and sustainable energy reviews*, *14*(8), 2434-2442.

[3] Katal, A., Dahiya, S., & Choudhury, T. (2022). Energy efficiency in the cloud computing data centers: a survey on software technologies. *Cluster Computing*, 1-31.

[4] Zhang, Q., Meng, Z., Hong, X., Zhan, Y., Liu, J., Dong, J. & Deen, M. J. (2021). A survey on data center cooling systems: Technology, power consumption modeling, and control strategy optimization. *Journal of Systems Architecture*, *119*, 102253.

[5] Murugesan, S. (2008). Harnessing Green IT: Principles and Practices. , 24–33. doi:10.1109/mitp.2008.10

[6] Harmon, Robert R.; Auseklis, Nora (2009). [IEEE Technology - Portland, OR, USA (2009.08.2-2009.08.6)] PICMET '09 - 2009 Portland International Conference on Management of Engineering & Technology - Sustainable IT services: Assessing the impact of green computing practices., 1707 1717. doi:10.1109/PICMET.2009.5261969

[7] Saha, Biswajit. "Green computing." International Journal of Computer Trends and Technology (IJCTT) 14.2 (2014): 46-50 [kp9].

[8] Kalyani, Vijay Laxmi, Manisha Kumari Dudy, and Shikha Pareek. "Green energy: The need of the world." Journal of Management Engineering and Information Technology 2.5 (2015): 18-26.

[9] Majid, M. A. "Renewable energy for sustainable development in India: current status, prospects, challenges, employment, and investment opportunities." Energy, Sustainability and Society 10.1 (2020): 1-36 [kp11].

[10] Yue, W., Xingzhu, H., & Lin, W. (2011). The development research of green economic in capital cities in Shandong. Energy Procedia, 5, 130-134]

[11] Atrey, A., Jain, N., & Iyengar, N. (2013). A study on green cloud computing. *International Journal of Grid and Distributed Computing, 6*(6),93

[12] Green Cloud Computing Ms. HaripriyaManikantGavali ISSN: 2321-9653; IC Value: 45.98; SJ Impact Factor: 7.538 Volume 10 Issue IV Apr 2022]

[13] Meghashree, N., Girija, R., & Sumathi, D. (2019). Understanding Green IoT: Research Applications and Future Directions.

[14] Meiyou, D., & Ye, Y. (2022). Establishment of a big data evaluation model for green and sustainable development of enterprises. *Journal of King Saud University-Science, 34*(5), 102041.

[15] Yigitcanlar, T., Mehmood, R., & Corchado, J. M. (2021). Green Artificial intelligence: Towards an efficient, sustainable, and equitable technology for smart cities and futures. *Sustainability, 13*(16), 8952.

[16] Benamer, W. H., Elberkawi, E. K., Neihum, N. A., Anwiji, A. S., & Youns, M. A. (2021). Green Computing case study: calls for proposing solutions for the Arabian Gulf Oil Company. In *E3S Web of Conferences* (Vol. 229, p. 01063). EDP Sciences.

[17] Prasad, S. S., Mishra, J. P., & Mishra, S. K. (2019, February). Analyzing Green Computation Linked to Cloud Resources: A Case Study. In *International Conference on Artificial Intelligence in Manufacturing & Renewable Energy (ICAIMRE).*

[18] Hamdan, M., Salem, A., & Shamayleh, Y. (2021, April). Harmonization between Renewable Energy and Cloud Computing towards Green Computing A Case Study: Data Center at The University Of Jordan. In *2021 12th International Renewable Engineering Conference (IREC)* (pp.1-5). IEEE.

[19] Shabeer, H. Abdul, et al. "Green cloud computing and communication." *Mobile Networks and Applications* 25 (2020): 1287-1289.

[20] Maksimovic, M. "Greening the future: Green Internet of Things (G-IoT) as a key technological enabler of sustainable development." Internet of things and big data analytics toward next-generation intelligence (2018): 283-313.

[21] Murugesan, San. "Harnessing green IT: Principles and practices." *IT* professional 10.1 (2008).

[22] Goodin, Dan. "IT confronts the data center power crisis." Infoworld, October (2006).

[23] Niloofar Gholipour, Ehsan Arianyan, Rajkumar Buyya, A Novel Energy-aware resource management technique using joint VM and

container consolidation approach for green computing in cloud data centers, Simulation Modelling Practice and Theory (2020), doi: https://doi.org/10.1016/j.simpat.2020.102127

[24] Jaiswal, A., Kumar, S., Kaiwartya, O., Prasad, M., Kumar, N. and Song, H., 2021. Green computing in IoT: Time-slotted simultaneous wireless information and power transfer. Computer Communications, 168, pp.155-169.

[25] Guangjie Han, Wenhui Que, Gangyong Jia, and Wenbo Zhang, Resource- utilization-aware energy-efficient server consolidation algorithm for green computing in IIOT, Journal of Network and Computer Applications, http://dx.doi.org/10.1016/j.jnca.2017.07.011

[26] Klimova A, Rondeau E, Andersson K, Porras J, Rybin A, Zaslavsky A, An International Master's program in green ICT as a contribution to sustainable development, Journal of Cleaner Production (2016), doi: 10.1016/j.jclepro.2016.06.032.

[27] Sakshi Student Member, IEEE, Rakesh Kumar Jha Senior Member, IEEE, Sanjeev Jain Senior Member, IEEE, A Comprehensive Survey on Green ICT with 5G-NB-IoT: Towards Sustainable Planet, Computer Networks (2021), doi: https://doi.org/10.1016/j.comnet.2021.108433

[28] Surah A. Bello, Automation in Construction, [2020] doi.org/10.1016/j.autcon.2020.103441

[29] Ankita Choudhary, Shilpa Ranab, K.J. Matahaic, "A Critical Analysis of Energy Efficient Virtual Machine Placement Techniques and its Optimization in a Cloud Computing Environment" (2015), doi: https://doi.org/10.1016/j.procs.2016.02.022

[30] Zoha Usmani, Shailendra Singh, "A Survey of Virtual Machine Placement Techniques in a Cloud Data Center", doi: https://doi.org/10.1016/j.procs.2016.02.093

[31] A Thakkar, K Chaudhari, M Shah, "A comprehensive survey on energy efficient power management techniques", doi: https://doi.org/10.1016/j.procs.2020.03.432

[32] Biswajit Debnath, Reshma Roychoudhuri, Sadhan K. Ghosh, "E-Waste Management – A Potential Route to Green Computing". doi: https://doi.org/10.1016/j.proenv.2016.07.063

[33] Monika Singh, Ganesh Prasad Sahu, "Towards adoption of Green IS: A literature review using classification methodology(2020) doi: https://doi.org/10.1016/j.ijinfomgt.2020.102147

[34] Adedapo Oluwaseyi Ojo, Murali Raman, Alan G. Downe, "Toward green computing practices: A Malaysian study of green belief and

attitude among Information Technology professionals" (2019) doi: https://doi.org/10.1016/j.jclepro.2019.03.237

[35] Yixiong Feng, Zhaoxi Hong, Zhiwu Li, Hao Zheng, Jianrong Tan, "Integrated intelligent green scheduling of sustainable flexible workshop with edge computing considering uncertain machine state" (2019) doi: https://doi.org/10.1016/j.jclepro.2019.119070

[36] André L.A. Di Salvo, Feni Agostinho, Cecília M.V.B. Almeida, Biagio F. Giannetti, "Can cloud computing be labeled as "green" Insights under an environmental accounting perspective" (2016) doi: https://doi.org/10.1016/j.rser.2016.11.153

[37] Farhan, L .Kharel, R., Kaiwartya, O., Hammoudeh, M., Adebisi, B., 2018. Towards green computing for Internet of things: Energy oriented path and message scheduling approach. Sustainable Cities and Society, 38, pp. 195-204. [pii: S2210670717309678]

[38] Huigui Rong , Haomin Zhang , Sheng Xiao , Canbing Li , Chunhua Hu, "Optimizing energy consumption for data centers" (2016), doi: https://doi.org/10.1016/j.rser.2015.12.283

[39] Al-Zamil A, Jilani Saudagar AK, "Drivers and Challenges of Applying Green Computing for Sustainable Agriculture: A Case Study", Sustainable Computing: Informatics and Systems (2018), https://doi.org/10.1016/j.suscom.2018.07.008

[40] Keke Gai, Meikang Qiu , Hui Zhao, Lixin Tao , Ziliang Zong, "Dynamic energy-aware cloudlet-based mobile cloud computing model for green computing" (2016) doi: https://doi.org/10.1016/j.jnca.2015.05.016

[41] Mirsaeid Hosseini Shirvani, Amir Masoud Rahmani, Amir Sahafi, "A survey study on virtual machine migration and server consolidation techniques in DVFS-enabled cloud datacenter: Taxonomy and challenges" (2018) doi: https://doi.org/10.1016/j.jksuci.2018.07.001

[42] I. Mohiuddin, A. Almogren, Workload aware VM consolidation method in edge/cloud computing for IoT applications, J. Parallel Distrib. Comput. (2018), https://doi.org/10.1016/j.jpdc.2018.09.011

[43] Wanneng Shu, Ken Cai, Neal Naixue Xiong, "Research on strong agile response task scheduling optimization enhancement with optimal resource usage in green cloud computing" (2021) doi: https://doi.org/10.1016/j.future.2021.05.012

Biographies

Sunil Kumar Pandey is currently working as a professor at the Institute of Technology and Science with an experience of over 24 years in industry and academia and have interest in cloud, blockchain, database technologies, and soft computing. Dr. Pandey has been credited with 65+ research papers (including SCI/Scopus Indexed), 3 book chapters, and 3 books with reputed publishers including Springer, IGI, IEEE Xplore, River Press – Denmark, Wiley, Hindawi, and journals/conferences. He has been a regular author of articles in different print and online platforms including interviews, views and has published 11 edited volumes on different relevant themes of information technologies. He has been providing and coordinating training and consultancy to various reputed organizations including the Indian Air Force and has conducted 25+ national/ international conferences/summits/conclaves in association with AICTE, CSI, DST, and other leading organizations. He has also conducted a large number of FDP/entrepreneurship programs supported by AICTE/DST/UGC/EDI, etc.

Kumar Pal Singh has 20 years of rich experience, in academia and industry. He is currently associated with the Institute of Technology and Science, Ghaziabad as an associate professor in the Department of IT. Mr. Singh did MCA and M.Tech. (CSE) and he is currently pursuing his Ph.D. in Computer Science and Engineering. The research area of Mr. Singh is cloud computing including fog and edge computing, and green computing. He has published more than 20 research papers, book chapters, and articles in various reputed international journals like IEEE Xplore, Springer, IJCSI, etc., and international/national conferences organized by reputed institutes and universities including IIT and NITs. He has attended various summits and technical programs organized by well-reputed corporate organizations including Google, Microsoft, Wipro, IBM, etc.

Puja Dhar, having an experience of over 18 years in academia, is presently working as an assistant professor and chairperson in MCA Program at I.T.S, Mohan Nagar, Ghaziabad. Ms. Dhar did her M.Sc. (IT) and M.Tech. (IT), and she is pursuing her Ph.D. in computer science and engineering. Her research area is machine learning and artificial intelligence. Her area of interest also includes database management systems, software engineering, cybersecurity, artificial intelligence, and green computing. She has published several research papers in Scopus-indexed journals and international conferences of repute.

Saurabh Saxena, MCA, M.Tech. (IT), is associated with the Department of IT, Institute of Technology and Science (affiliated to AKTU, Lucknow), Mohan Nagar, Ghaziabad, India. He has more than 18 years of academic experience. He is an alumnus of GGSIP University, Delhi. His area of expertise is software testing, software engineering, database management system, green computing, and data mining and warehousing. He has been invited as a resource person in various workshops and invited to talk at the institute and university levels. He has been awarded three times for his outstanding performance in teaching at his present working organization. He has published 20+ research papers in international journal and conference proceedings. Currently, he is pursuing his Ph.D. from JIIT Noida

Kumar Vaibhav Bhatnagar has over nine years of experience in academia and industry. He is currently working as an assistant professor at the Institute of Technology and Science, Ghaziabad in the Department of IT. Mr. Bhatnagar did his M.Tech. (WSN) and B.E. (ECE), and he is currently pursuing his Ph.D. in computer science and engineering with a research area in cybersecurity for IoT. He has been into computer networks having certifications of Cisco Network Administrator and networking diploma.

9

Artificial Intelligence and Green 6G Network-enabled Architectures, Scenarios, and Applications for Autonomous Connected Vehicles

Sachin Sharma[1], Ranu Tyagi[1], and Seshadri Mohan[2]

[1]Graphic Era Deemed to be University, India
[2]University of Arkansas at Little Rock, USA

Abstract

The implementation of green 6G networks by the telecommunications industry will spur further innovations and speed up the wireless technology evolution. The expansion and application of these networks to autonomous connected vehicles (ACV) will influence the spectrum bidding policy for green 6G networks and its implementation. The objectives and expectations of 6G network establishment will be orders of magnitude more than those of 5G network communication requirements, such as ultra-high energy capacity, ultra-high network throughput, ultra-high reliability, ultra-high performance, excellent Quality of Service (QoS), massive ultra-low latency, and ultra-high customer satisfaction ratio. With the goal of the aforementioned expectations, a significant amount of infrastructural changes or updates will be required in the established implementation of wireless or mobile communication networks. Efforts are underway worldwide toward planning for 6G that will pave the way for the adoption of green 6G networks for future ACVs by 2035. These networks will most certainly exploit the capabilities of artificial intelligence (AI) and machine intelligence to provide large number innovative applications at low communication overhead costs, with efficient surveillance, security, and privacy solutions to commercial and domestic users. In this chapter, we will examine the green 6G network-enabled architectures, scenarios, challenges, and Internet of Vehicles (IoV) applications for future ACVs. We also discuss the applicability of AI and machine learning (ML) paradigm for ACVs through several use cases.

Keywords: Green 6G Network, Internet of Vehicles, Artificial Intelligence, Machine Learning, Autonomous Connected Vehicles

9.1 Introduction

Green 6G network-enabled architectures and artificial intelligence (AI) are poised to transform how we perceive linked autonomous vehicles. The term artificial intelligence (AI) describes a machine's capacity for functions like learning, thinking, and decision-making that typically needs human intelligence [1]. On the other hand, green 6G networks are the newest generation of mobile communication systems that are intended to be more sustainable and energy-efficient than their forerunners. Together, these technologies have the potential to open up new avenues for linked autonomous vehicles, allowing them to function more safely and effectively than before [2]. Ability to recognize and react to real-time driving conditions, such as traffic jams, road dangers, and shifting weather patterns, is a significant use of AI in autonomous vehicles. AI algorithms can assist vehicles in adapting to changing situations by analyzing enormous volumes of data in real time, lowering the chance of accidents and enhancing overall performance [3]. This kind of real-time communication between cars and their surroundings will be made possible by green 6G networks. Vehicles will be able to interact in real time with other vehicles as well as with other infrastructure, such as traffic lights and road signs, thanks to 6G networks' ultra-fast and dependable communication. Vehicles will be able to better coordinate their movements as a result, which will ease traffic congestion and enhance overall traffic flow. Analysis and optimization of energy use is another significant use of AI in connected vehicles. AI algorithms can help automobiles function more effectively, cutting down on energy use and emissions, by monitoring variables like battery life, engine efficiency, and environmental conditions. With the high-speed, low-latency connectivity required to support real-time energy optimization, green 6G networks will play a crucial role in assisting these kinds of energy-efficient applications [4]. For autonomous connected vehicles, the combination of AI and green 6G network-enabled architectures represents a significant advancement. These technologies will open up new opportunities for safety, efficiency, and sustainability by enabling vehicles to communicate and adapt in real time, paving the way for a more connected and sustainable future.

9.2 Overview of Autonomous Connected Vehicles

A new generation of vehicles called autonomous connected vehicles use cutting-edge technology to offer safer, more effective, and more convenient transportation. Autonomous vehicles, also referred to as self-driving automobiles, function devoid of human oversight thanks to a combination of sensors, software, and AI algorithms [5]. These automobiles can operate autonomously, sensing their surroundings, traversing the roadways, and making judgments in the present. On the other hand, connected automobiles make use of cutting-edge communication technologies to exchange data with infrastructure, networks, and other vehicles. Connected vehicles can deliver real-time updates on traffic situations, weather, and road dangers by employing sensors and communication systems to gather and communicate information. This enables drivers to make better judgments and prevent accidents. The fusion of autonomous and connected technology is poised to transform how we view transportation. Autonomous vehicles have the ability to significantly reduce traffic fatalities and accidents while simultaneously enhancing traffic flow and consuming less energy. Real-time updates on the state of the roads can be provided by connected vehicles, helping drivers avoid traffic jams and collisions while also enhancing overall safety [6]. Autonomous connected vehicles provide a variety of new mobility options in addition to these advantages. For instance, on-demand transport services using autonomous vehicles might do away with the necessity for automobile ownership and lower the number of vehicles on the road. New business models, like mobile offices or entertainment hubs, might be supported by connected vehicles, giving passengers new ways to work and play while traveling. These innovations have the power to revolutionize how we travel, making it safer, more effective, and more convenient than before.

9.3 Emerging Technologies in Autonomous Connected Vehicles

A number of cutting-edge technologies are poised to be essential in the creation and adoption of autonomous linked vehicles. A few of the important technologies are shown in Figure 9.1.

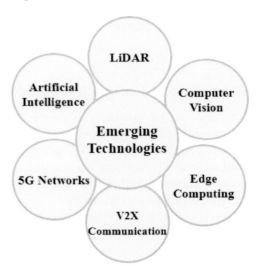

Figure 9.1   Emerging technologies in autonomous connected vehicles.

- *LiDAR:* LiDAR is a remote sensing technology that uses laser beams to measure distances and generates 3D maps of the surrounding environment. LiDAR sensors are commonly used in autonomous vehicles to provide accurate and real-time data on the vehicle's surroundings, allowing it to navigate roads and avoid obstacles.
- *Computer vision:* Computer vision is a technology that uses cameras and AI algorithms to interpret and analyze visual data. In autonomous vehicles, computer vision is used to identify objects and pedestrians, detect traffic signs and signals, and recognize road markings.
- *Edge computing:* Edge computing refers to the ability to process data in real time at the edge of a network, rather than sending it to a centralized data center. In autonomous connected vehicles, edge computing is used to process and analyze data in real time, allowing vehicles to make decisions quickly and efficiently.
- *V2X communication:* Vehicle-to-everything (V2X) communication is a technology that allows vehicles to communicate with other vehicles, infrastructure, and networks. V2X communication is critical for enabling connected vehicles to exchange real-time data on traffic conditions, weather, and road hazards [7].

- *5G networks:* 5G networks are the latest generation of mobile communication systems, offering faster speeds, lower latency, and higher reliability than previous generations. 5G networks are critical for enabling real-time communication between autonomous connected vehicles and their surrounding environment.
- *Artificial intelligence:* AI algorithms are used in autonomous connected vehicles to analyze and interpret vast amounts of data in real time. AI enables vehicles to make decisions and adapt to changing conditions quickly and efficiently, improving safety and efficiency.

9.4 AI and Green 6G Network-enabled Architectures

Autonomous connected car development and implementation depend heavily on AI and green 6G network-enabled architectures. Through real-time communication and data interchange made possible by these technologies, vehicles may provide precise and current information on traffic situations, road hazards, and weather trends. They also make it possible for automobiles to run more effectively, which lowers energy use and lowers carbon emissions [8]. The application of machine learning algorithms is a crucial component of AI-enabled infrastructures. Large-scale real-time data analysis performed by machine learning algorithms can reveal patterns and trends that can be used to enhance the performance of networked autonomous cars. Machine learning algorithms, for instance, can be used to forecast traffic patterns and modify a vehicle's route to avoid congestion, cutting down on travel time and energy use. The application of speech recognition and natural language processing (NLP) technologies is a key component of AI-enabled infrastructures. Using voice instructions and everyday language, NLP and speech recognition technologies allow drivers and passengers to engage with autonomous connected vehicles in a natural and intuitive way [9]. Autonomous connected car development also depends on green 6G network-enabled designs. These architectural designs make use of cutting-edge communication technologies to deliver up-to-the-minute information about traffic, weather, and road dangers. Additionally, they allow for communication between cars, networks, and infrastructure, resulting in more effective and well-coordinated transportation systems. Autonomous connected vehicles can operate more effectively and sustainably than ever before by utilizing AI and green 6G network-enabled infrastructures. They can also offer a variety of fresh travel options, such as on-demand services, mobile workplaces, and entertainment hubs.

9.5 AI and Green 6G Network-enabled Scenarios

The way we think about transport is poised to change as a result of a number of scenarios for autonomous connected vehicles that are enabled by AI and a green 6G network. Several of the important cases include (Figure 9.2):

- *Autonomous on-demand transportation services:* AI and green 6G network-enabled architectures will enable the development of autonomous on-demand transportation services. These services will allow passengers to summon a vehicle using a smartphone app, and the vehicle will arrive at their location without a driver. By leveraging AI and green 6G network-enabled architectures, these vehicles can operate more efficiently, reducing energy consumption and minimizing carbon emissions.

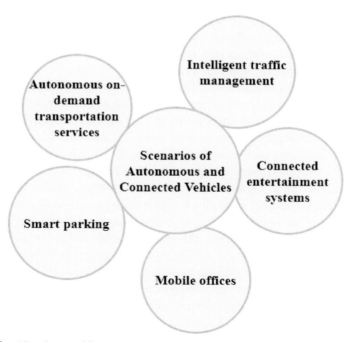

Figure 9.2   AI and green 6G network-enabled scenarios for autonomous connected vehicles.

- *Intelligent traffic management:* AI and green 6G network-enabled architectures can be used to develop intelligent traffic management systems that can optimize traffic flow and reduce congestion. By analyzing real-time data on traffic patterns and road conditions, these systems can adjust traffic signals and redirect vehicles to less congested routes, reducing travel time and energy consumption [10].

- *Connected entertainment systems:* AI and green 6G network-enabled architectures can enable the development of connected entertainment systems for autonomous connected vehicles. These systems can provide passengers with access to a range of entertainment options, including streaming video and music services, gaming platforms, and virtual reality experiences, making the travel experience more enjoyable and engaging.

- *Mobile offices:* AI and green 6G network-enabled architectures can enable the development of mobile offices for autonomous connected vehicles. These systems can provide passengers with access to high-speed Internet, productivity tools, and communication platforms, allowing them to work while on the move.

- *Smart parking:* AI and green 6G network-enabled architectures can be used to develop smart parking systems that can optimize parking availability and reduce congestion. By analyzing real-time data on parking availability and vehicle locations, these systems can direct vehicles to available parking spaces, reducing the time and energy spent searching for parking.

9.6 Applications

Numerous AI applications and architectures for autonomous connected vehicles supported by green 6G networks are anticipated to revolutionize the transportation sector. A few of the important applications include (Figure 9.3):

- *Intelligent navigation:* Intelligent navigation systems can be made possible by AI and green 6G network topologies, which can optimize route selection and reduce overall journey time. These technologies are capable of analyzing real-time traffic information and offering drivers route recommendations that minimize congestion and trip time.

- *Predictive maintenance:* AI and green 6G network-enabled architectures can enable predictive maintenance systems that can detect potential

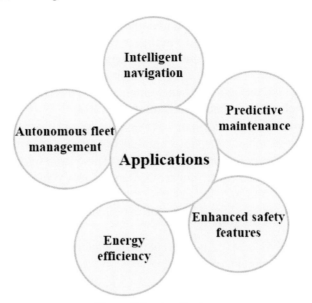

Figure 9.3   Applications.

problems with autonomous connected vehicles before they become major issues. These systems can analyze real-time vehicle data and identify potential issues, allowing for proactive maintenance and repair.

- *Enhanced safety features:* AI and green 6G network-enabled architectures can enable enhanced safety features for autonomous connected vehicles. These features can include advanced driver assistance systems, collision avoidance systems, and emergency braking systems, all of which can improve overall safety on the roads.

- *Energy efficiency:* AI and green 6G network-enabled architectures can enable more energy-efficient operation of autonomous connected vehicles. These systems can optimize vehicle performance based on real-time data, reducing energy consumption and minimizing carbon emissions.

- *Autonomous fleet management:* AI and green 6G network-enabled architectures can enable the development of autonomous fleet management systems, which can optimize the use of autonomous vehicles in a fleet. These systems can analyze real-time data on vehicle location, passenger demand, and traffic conditions, allowing for more efficient deployment of vehicles and reduced wait times for passengers [11].

9.7 Open Source Tools for Autonomous Connected Vehicles

For the creation of autonomous linked automobiles, a number of open source technologies are available. These tools offer programmers a variety of resources for creating and testing connected and autonomous car technologies. Among the most important open source resources for linked autonomous vehicles are (Figure 9.4):

- *Autoware:* Autoware is an open source software platform for autonomous vehicles. It provides a range of tools for autonomous driving, including sensor fusion, localization, and mapping.
- *Apollo:* Apollo is an open source software platform for autonomous vehicles that was developed by Baidu. It provides a range of tools for autonomous driving, including perception, planning, and control.
- *ROS (Robot operating system):* ROS is an open source framework for robotics development. It provides a range of tools and libraries for developing autonomous and connected vehicle technologies, including sensor data processing, mapping, and navigation.
- *Open CV (Open source computer vision):* Open CV is an open source library for computer vision development. It provides a range of tools for processing and analyzing visual data, which is critical for developing autonomous and connected vehicle technologies.

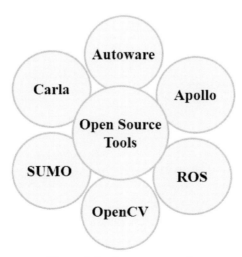

Figure 9.4 Open source tools.

- *SUMO (Simulation of urban mobility):* SUMO is an open source traffic simulation software that can be used to test and validate autonomous and connected vehicle technologies. It provides a range of features for simulating traffic flow and vehicle behavior, which can be used to test and refine autonomous vehicle algorithms.
- *Carla:* Carla is an open source simulator for autonomous driving research. It provides a high-fidelity simulation environment for testing autonomous vehicle algorithms, including sensor data processing, perception, and control.

9.8 Advantages

AI and green 6G network-enabled designs for autonomous connected vehicles have a number of benefits, including (Figure 9.5):

- *Improved safety:* AI and green 6G network-enabled architectures can enable enhanced safety features for autonomous connected vehicles, including collision avoidance systems, emergency braking systems, and advanced driver assistance systems. These features can significantly reduce the risk of accidents on the roads [12].

Figure 9.5 Advantages.

- *Increased efficiency:* AI and green 6G network-enabled architectures can optimize the performance of autonomous connected vehicles, reducing travel times and increasing overall efficiency. These systems can analyze real-time data on traffic, road conditions, and passenger demand, allowing for more efficient deployment of vehicles and reduced wait times for passengers.

- *Enhanced sustainability:* AI and green 6G network-enabled architectures can enable more sustainable transportation systems by reducing carbon emissions and minimizing energy consumption. These systems can optimize vehicle performance based on real-time data, reducing energy consumption, and minimizing the environmental impact of transportation [13].

- *Improved user experience:* AI and green 6G network-enabled architectures can enhance the user experience for passengers by providing more personalized and efficient transportation services. These systems can analyze passenger data, such as location and destination, to provide customized route suggestions and improve overall travel experience [14].

- *Cost savings:* AI and green 6G network-enabled architectures can enable cost savings for transportation providers by optimizing vehicle performance and reducing maintenance costs. Predictive maintenance systems can identify potential issues before they become major problems, reducing repair costs and downtime for vehicles.

9.9 Challenges

While AI and green 6G network-enabled designs for autonomous connected vehicles have numerous benefits, there are also a number of issues that need to be resolved. Some of the major difficulties include (Figure 9.6):

- *Privacy and security concerns:* As autonomous connected vehicles collect and transmit large amounts of data, there are concerns about the privacy and security of this data. Ensuring the protection of sensitive data will be critical for the successful adoption of these technologies [15].

- *Infrastructure requirements:* AI and green 6G network-enabled architectures will require significant infrastructure investments to support the deployment of autonomous connected vehicles. This includes the development of 5G and 6G networks, as well as the installation of sensors and other equipment required for vehicle connectivity.

Figure 9.6   Challenges.

- *Legal and regulatory challenges:* The deployment of autonomous connected vehicles will require new legal and regulatory frameworks to ensure the safety and security of these systems. This includes the development of new laws and regulations related to data privacy, cybersecurity, and liability for accidents involving autonomous vehicles.
- *Technical challenges:* The development of AI and green 6G network-enabled architectures for autonomous connected vehicles requires overcoming technical challenges related to data processing, machine learning algorithms, and real-time communication between vehicles and infrastructure [16].
- *Public acceptance:* The successful deployment of autonomous connected vehicles will depend on public acceptance and trust in these technologies. This will require education and awareness campaigns to help people understand the benefits and risks of these technologies.

9.10 Case Studies of AI and Green 6G Network-enabled Architectures for Autonomous Connected Vehicles

The government and business sectors have investigated a number of AI and green 6G network-enabled architectures for autonomous connected vehicle-related projects in various nations.

- *VENTURER project, UK:* VENTURER project, which was a UK-based initiative aimed at exploring the technical, social, and legal implications of autonomous vehicles. As part of the VENTURER project, a fleet of autonomous vehicles was deployed on public roads in Bristol, UK. The vehicles were equipped with advanced AI and green 6G network-enabled architectures, including sensors and communication systems that enabled real-time data analysis and communication with other vehicles and infrastructure. The project focused on several key areas, including safety, user experience, and environmental sustainability. For example, the autonomous vehicles were equipped with advanced safety features, such as collision avoidance systems and emergency braking, to minimize the risk of accidents on the roads. The vehicles were also designed to optimize their performance based on real-time data, reducing energy consumption and minimizing environmental impact. To ensure the successful deployment of the autonomous vehicles, the VENTURER project also addressed several challenges related to infrastructure, legal and regulatory frameworks, and public acceptance. This included the development of new regulations related to autonomous vehicles and the establishment of public engagement programs to educate the public about the benefits and risks of these technologies. The VENTURER project demonstrated the potential of AI and green 6G network-enabled architectures for autonomous connected vehicles. By addressing technical, social, and legal challenges, the project provided valuable insights into the development of next-generation transportation systems. The success of the VENTURER project highlights the potential of AI and green 6G network-enabled architectures for improving safety, efficiency, and sustainability in the transportation industry.

- *Smart Mobility Corridor in Ohio, USA:* Currently being created is a 35-mile length of highway for testing and deploying cutting-edge transportation technology, such as networked driverless vehicles, smart infrastructure, and communication systems. Fiber optic cable, roadside sensors, and other communication infrastructure are all included in the Smart Mobility Corridor, allowing for real-time data sharing between traffic control systems, infrastructure, and automobiles. The project aims to increase mobility for locals and companies while reducing traffic congestion and improving transit safety. The utilization of linked car technology, which enables vehicles to communicate with each other and with infrastructure in real time, is one of the essential aspects of the Smart Mobility Corridor. This technology enables more effective traffic management and routing by warning vehicles of impending risks like accidents or road closures. Utilizing cutting-edge analytics and machine learning algorithms to analyze data from sensors and other sources is another crucial component of the Smart Mobility Corridor. Utilizing this information will increase energy efficiency, traffic flow, and overall transportation effectiveness. The creation of environmentally friendly transportation options, such as electric and hybrid vehicles, is another goal of the Smart Mobility Corridor. Along the corridor, electric vehicle charging stations will be installed, and new battery and charging infrastructure will also be tested. The US's development of AI and green 6G network-enabled architectures for autonomous connected vehicles has advanced significantly thanks to the Smart Mobility Corridor. The initiative is assisting in the development of a more effective, egalitarian, and sustainable transportation system for local citizens and companies by utilizing cutting-edge communication and analytics technologies.

- *Grand Field Ota project, Ota City, Tokyo, Japan:* By combining linked autonomous vehicles with smart infrastructure and renewable energy sources, this initiative intends to develop a sustainable transportation system. The Grand Field Ota project calls for the creation of an electric, self-driving vehicle that can travel a predetermined route without human intervention. The bus has sensors and cameras that allow it to recognize impediments and modify its speed and path as necessary. Additionally, it has real-time communication with infrastructure, other vehicles, and both, enhancing traffic management and safety. The Grand Field Ota project includes the implementation of smart infrastructure, like traffic lights and road sensors that can communicate with autonomous connected cars in addition to the self-driving bus. To cut carbon emissions and encourage sustainability, this infrastructure is also fueled by renewable energy sources like solar and wind energy. In order to improve traffic flow and lower energy usage, the project also makes use of machine learning and advanced analytics. For instance, to ease congestion and boost overall traffic efficiency, the system can analyze traffic patterns and modify the timing of traffic lights. The Grand Field Ota project shows how green 6G networks and AI-enabled architectures may be used to build a reliable and effective transport system. The project is assisting in lowering carbon emissions, improving mobility, and enhancing the quality of life for citizens in Ota City and beyond by combining autonomous linked vehicles with smart infrastructure and green energy sources.

- *K-City project, South Korea:* It is a large-scale testbed for autonomous connected vehicles located in Hwaseong, Gyeonggi Province. The K-City project covers an area of 320,000 square meters and includes a variety of urban and suburban driving scenarios, such as intersections, tunnels, and parking lots, to simulate real-world driving conditions. The project also includes various types of smart infrastructure, such as traffic lights, road sensors, and cameras, which enable real-time communication between vehicles and infrastructure. The K-City project is powered by a 5G network that enables high-speed data exchange between vehicles and infrastructure, as well as real-time data analysis and decision-making. The project also includes the use of advanced artificial intelligence and machine learning algorithms to optimize traffic flow, reduce energy consumption, and enhance safety. One of the key features of the K-City project is the use of autonomous shuttle buses, which provide transportation for visitors and employees within the testbed. These shuttle buses are equipped with sensors and cameras that enable them to operate autonomously, and they communicate with other vehicles and infrastructure in real time to avoid collisions and optimize traffic flow. The K-City project also includes the use of green transportation solutions, such as electric and hydrogen fuel cell vehicles, which help to reduce carbon emissions and promote sustainability. The K-City project demonstrates the potential of AI and green 6G network-enabled architectures for autonomous connected vehicles in creating a more efficient, sustainable, and safe transportation system. By leveraging advanced communication, analytics, and green energy technologies, the project is helping to pave the way for the future of transportation in South Korea and beyond.

- *Intelligent and Green Mobility project, Tianjin, China:* This project aims to create a smart transportation system that integrates autonomous connected vehicles with green energy sources and intelligent infrastructure. The project includes the deployment of autonomous shuttles, which run on electricity and can operate autonomously on designated routes. These shuttles are equipped with sensors and cameras that enable them to detect obstacles and adjust their speed and route accordingly. They also communicate with other vehicles and infrastructure in real time, enabling more efficient traffic management and safety. In addition to the autonomous shuttles, the Intelligent and Green Mobility project includes the use of smart infrastructure, such as traffic lights and road sensors that can communicate with autonomous connected vehicles. This infrastructure is also powered by renewable energy sources, such as solar panels and wind turbines, to reduce carbon emissions and promote sustainability. The project also includes the use of advanced analytics and machine learning algorithms to optimize traffic flow and reduce energy consumption. For example, the system can analyze traffic patterns and adjust traffic light timings to reduce congestion and improve overall traffic efficiency. A 5G network is also incorporated into the Intelligent and Green Mobility project, allowing for high-speed data interchange between vehicles and infrastructure as well as real-time data analysis and decision-making. In order to increase user experience and safety, this network also facilitates the adoption of other cutting-edge technologies like augmented reality and virtual reality. The Intelligent and Green Mobility project shows how AI and environmentally friendly 6G network-enabled architectures can be used to develop a more effective, intelligent, and sustainable transport system in China and beyond. The initiative is assisting in lowering carbon emissions, improving transportation, and enhancing the quality of life for citizens in Tianjin and beyond by integrating autonomous connected vehicles with smart infrastructure and green energy sources.

- *Connected Autonomous Vehicles (CAV) program, Victoria, Australia:* This program is a partnership between the Victorian Government, Telstra, and Lexus Australia, with the aim of exploring the potential of autonomous connected vehicles in improving road safety and reducing congestion. The CAV program includes the deployment of autonomous connected vehicles on selected roads and highways in Victoria, equipped with advanced sensors and communication technologies. The vehicles are able to communicate with other vehicles and infrastructure in real time, enabling better traffic management and coordination. The program also incorporates the use of advanced analytics and machine learning algorithms to optimize traffic flow and improve safety. For example, the system can analyze traffic patterns and adjust traffic light timings to reduce congestion and improve overall traffic efficiency. In addition to the autonomous vehicles, the CAV program includes the use of smart infrastructure such as traffic lights, road sensors, and parking lots that can communicate with autonomous connected vehicles. This infrastructure is powered by renewable energy sources such as solar panels and wind turbines, to reduce carbon emissions and promote sustainability. The CAV program also leverages the 6G network for high-speed data exchange between vehicles and infrastructure, as well as real-time data analysis and decision-making. This network also supports the deployment of other emerging technologies, such as augmented reality and virtual reality, to enhance the user experience and improve safety. The CAV program demonstrates the potential of AI and green 6G network-enabled architectures for creating a safer, more efficient, and sustainable transportation system in Australia and beyond. By integrating autonomous connected vehicles with smart infrastructure and green energy sources, the program is helping to reduce carbon emissions, improve mobility, and enhance the quality of life for residents in Victoria and beyond.

- *Smart Mobility Africa, South Africa:* The African Development Bank and the United Nations Economic Commission for Africa (UNECA) have joined forces to launch it. The project intends to provide a comprehensive framework for smart mobility in African nations, which includes the usage of connected autonomous vehicles, smart infrastructure development, and renewable energy sources. The initiative focuses on numerous important areas, including mobility for people with disabilities, logistics, and public transit. Promoting sustainability through lowering carbon emissions and increasing the usage of renewable energy sources is one of the project's main goals. The project uses renewable energy to power the infrastructure and cars in order to meet this objective. Examples of these sources are solar and wind power. The project also makes use of AI and machine learning algorithms to streamline travel plans, ease traffic, and boost security. For instance, to reduce delays and boost efficiency, the system can analyze traffic patterns and real-time modify transportation timetables. The "City Intelligence and Mobility Solutions" project in Cape Town, South Africa, is another initiative in Africa. The project, a partnership between the municipal administration, regional academic institutions, and tech firms, aims to create AI and environmentally friendly 6G network-enabled architectures for autonomous connected vehicles. As part of the project, driverless shuttles will be used for public transit, smart traffic management systems will be created, and renewable energy sources will be used to power the infrastructure. The initiative also uses AI and machine learning algorithms to streamline travel plans, ease traffic, and improve user experience. These initiatives show the potential of green 6G networks and AI-enabled architectures for connected autonomous vehicles in African nations. These initiatives contribute to building a more connected, accessible and sustainable transport system for the continent by encouraging sustainability, increasing efficiency, and improving the user experience.

- *Smart Mobility project, Santiago, Chile, South America:* With the use of green energy sources, sophisticated infrastructure, and networked autonomous vehicles, this initiative intends to develop an integrated transportation system. The Smart Mobility project calls for the use of electric autonomous shuttles that travel along predetermined routes and are fitted with cameras and sensors that can detect obstructions and cause them to change their speed or course. Real-time communication between the shuttles and infrastructure, as well as other vehicles, makes for safer and more effective traffic management. The Smart Mobility initiative also uses smart infrastructure, such as traffic signals and road sensors, which can talk to connected autonomous vehicles in addition to the autonomous shuttles. To cut carbon emissions and advance sustainability, this infrastructure is fueled by renewable energy sources like solar panels and wind turbines. To improve traffic flow and lower energy usage, the project also incorporates sophisticated analytics and machine learning techniques. For instance, to ease congestion and boost overall traffic efficiency, the system can analyze traffic patterns and modify the timing of traffic lights. A 6G network is also a part of the Smart Mobility initiative, allowing for quick data interchange between vehicles and infrastructure as well as real-time data analysis and decision-making. In order to increase user experience and safety, this network also facilitates the adoption of other cutting-edge technologies like augmented reality and virtual reality. The Smart Mobility project shows how green 6G networks and AI-enabled architectures may be used to build a more effective, intelligent, and sustainable transport system in South America and elsewhere. The project is assisting in lowering carbon emissions, improving transportation, and enhancing the quality of life for citizens in Santiago and beyond by combining autonomous linked vehicles with smart infrastructure and renewable energy sources.

- *Autonomous Lab project, Paris, France:* The goal of this project, a partnership between the Paris region, the city of Paris, and Group Renault, is to create and test innovative mobility services utilizing linked autonomous vehicles. In the Autonomous Lab project, electric autonomous vehicles will be used that will go along predetermined routes and be fitted with sensors, cameras, and other cutting-edge technologies for communication, perception, and navigation. Real-time communication between the vehicles and infrastructure allows for improved traffic management and safety. The Autonomous Lab project also uses smart infrastructure, such as traffic lights, road sensors, and parking lots, which can communicate with autonomous connected vehicles in addition to the autonomous vehicles. To cut carbon emissions and encourage sustainability, this infrastructure is fueled by renewable energy sources including solar and wind energy. To improve traffic flow and lower energy usage, the project also uses advanced analytics and machine learning algorithms. For instance, to ease congestion and boost overall traffic efficiency, the system can analyze traffic patterns and modify the timing of traffic lights. The Autonomous Lab project also makes use of the 6G network for real-time data analysis and decision-making, as well as for high-speed data interchange between vehicles and infrastructure. In order to improve user experience and safety, this network also supports the adoption of other cutting-edge technologies, such as augmented reality and virtual reality. The Autonomous Lab project shows the promise of green 6G networks and architectures enabled by AI for developing a more effective, intelligent, and sustainable transport system in France and beyond. The project is assisting in lowering carbon emissions, improving transportation, and enhancing the quality of life for citizens in Paris and the surrounding area by combining autonomous linked vehicles with smart infrastructure and renewable energy sources.

- *Dubai Autonomous Transportation Strategy, United Arab Emirates:* The strategy is a government-led initiative aimed at making 25% of all transportation trips in Dubai autonomous by 2030, with the ultimate goal of improving mobility, reducing congestion, and enhancing the sustainability of the transportation sector. The strategy includes several initiatives, such as the deployment of autonomous vehicles for public transportation, including taxis and buses. These vehicles are equipped with advanced sensors and communication technologies that enable them to navigate the city safely and efficiently, while communicating with other vehicles and infrastructure in real time. The strategy also incorporates the use of smart infrastructure, such as intelligent traffic management systems and smart parking, which can communicate with autonomous connected vehicles to optimize traffic flow and reduce congestion. Additionally, the infrastructure is powered by renewable energy sources such as solar panels, to promote sustainability and reduce carbon emissions. The Dubai Autonomous Transportation Strategy also leverages AI and machine learning algorithms to optimize transportation routes and schedules, as well as to enhance the user experience. For example, the system can analyze traffic patterns and user preferences to suggest the most efficient and comfortable route for a given trip. The strategy also includes the use of the 6G network for high-speed data exchange between vehicles and infrastructure, as well as real-time data analysis and decision-making. This network also supports the deployment of other emerging technologies, such as augmented reality and virtual reality, to enhance the user experience and improve safety. The Dubai Autonomous Transportation Strategy demonstrates the potential of AI and green 6G network-enabled architectures for creating a more efficient, sustainable, and user-friendly transportation system in Middle Eastern countries and beyond. By integrating autonomous connected vehicles with smart infrastructure and green energy sources, the strategy is helping to reduce carbon emissions, improve mobility, and enhance the quality of life for residents and visitors in Dubai and beyond.

- *Smart Transportation Innovation Cluster, Israel:* Several projects investigating the use of AI and green 6G network-enabled architectures for autonomous connected vehicles are based in Israel. The "Smart Transportation Innovation Cluster," a partnership between the government, academia, and business to advance the development of smart transportation solutions, is one such project. The cluster intends to create AI and machine learning algorithms that can optimize travel routes, ease traffic, and boost security. For instance, to reduce delays and boost efficiency, the system can analyze traffic patterns and real-time modify transportation timetables. The "Smart Mobility Initiative," a collaborative project of the Israeli government and the World Bank, is another program. The project's goal is to create a comprehensive framework for smart mobility in Israel, which will involve the usage of linked autonomous vehicles, smart infrastructure development, and renewable energy sources. The project uses renewable energy to power the infrastructure and cars in order to meet this objective. Examples of these sources are solar and wind power. The project also makes use of AI and machine learning algorithms to streamline travel plans, ease traffic, and boost security. Promoting sustainability through lowering carbon emissions and increasing the usage of renewable energy sources is one of the project's main goals. In order to do this, the project is investigating the usage of hybrid and electric cars, as well as the creation of battery-swapping stations and charging infrastructure. Additionally, there are a number of start-ups in Israel working on AI and green 6G network-enabled architectures for connected autonomous vehicles. For instance, the Israeli company "MobilEye," which Intel just acquired, is creating ADAS, or advanced driver assistance systems, to increase safety and decrease accidents. These initiatives show Israel's potential for autonomous connected car systems supported by green 6G networks and artificial intelligence. These initiatives contribute to the development of a more connected, accessible and sustainable transportation system for the nation by encouraging sustainability, increasing efficiency, and improving user experience.

- *Smart Cities Mission, India:* In order to meet the problems of its fast-expanding urban population and the resulting rise in demand for transportation, India is now researching the usage of AI and green 6G network-enabled architectures for autonomous linked vehicles. The Indian government's "Smart Cities Mission," which aims to create effective and sustainable urban transport systems by utilizing cutting-edge technologies like AI, IoT, and green 6G networks, is one example of this. As part of this strategy, a number of cities are creating pilot projects to test linked autonomous vehicles in actual traffic situations. Additionally, a number of Indian firms are creating architectures for autonomous connected vehicles that are supported by AI and green 6G networks. An Indian firm called "RideCell," for instance, is creating an intelligent platform that makes use of machine learning and AI to optimize travel routes and ease traffic. The platform also considers the development of battery-swapping stations and charging infrastructure, as well as green energy sources like electric and hybrid vehicles. Another

Indian firm, "Swiggy," is working on a drone- and autonomous-vehicle-based food delivery service. By leveraging AI and machine learning algorithms to optimize delivery routes and lessen congestion, the service seeks to increase efficiency and decrease delivery times. As part of its commitment to lowering carbon emissions and promoting sustainability, the Indian government is also encouraging the development of electric and hybrid vehicles. The government is offering incentives for the purchase of electric and hybrid vehicles as well as for the construction of charging infrastructure and battery-swapping stations under the "National Electric Mobility Mission Plan." The initiatives in India show the promise of green 6G networks and AI-enabled architectures for autonomous connected vehicles to meet urbanization's difficulties and advance sustainability. These projects are assisting in the development of a more connected, accessible, and sustainable transportation system for the nation by streamlining transportation routes, reducing traffic, and encouraging the use of green energy sources.

9.11 Future Perspectives

Future prospects for AI and eco-friendly 6G network-enabled autonomous linked car technologies look bright. Here are some potential changes to watch for: (Figure 9.7):

- *Increased adoption of autonomous connected vehicles:* In the upcoming years, we may anticipate seeing a wider acceptance of autonomous connected vehicles thanks to advancements in AI and eco-friendly 6G network architectures. As a result, transportation effectiveness, safety, and environmental sustainability will all increase [17].
- *Development of new business models:* The adoption of autonomous connected vehicles will probably result in the creation of new business models like shared mobility and mobility-as-a-service (MaaS). These models will make it possible to employ transport resources more effectively, resulting in less traffic congestion and better air quality [18].
- *Integration with smart cities:* Intelligent city infrastructure will be combined with autonomous connected vehicles, allowing for real-time data exchange and communication between vehicles, infrastructure, and pedestrians. As a result, metropolitan regions will see improved traffic flow, lower energy use, and higher general quality of life [19].

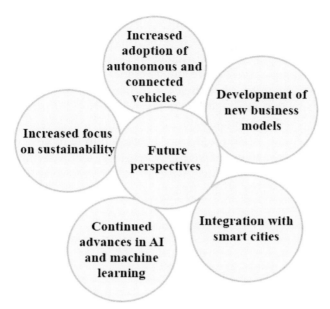

Figure 9.7 Future perspectives.

- *Continued advances in AI and machine learning:* We may anticipate ever more advanced autonomous and connected vehicle systems as AI and machine learning algorithms develop. These devices will be able to adjust to shifting road conditions and offer passengers individualized transit experiences [20].
- *Increased focus on sustainability:* We can anticipate a greater emphasis on creating green transport options due to growing concern over climate change and environmental sustainability. In order to create more environmentally friendly transportation systems, such electric and hybrid automobiles, and lower transportation-related carbon emissions, AI and green 6G network-enabled designs will be crucial.

9.12 Conclusion

By enabling the creation of autonomous connected vehicles that are safer, more effective, and more sustainable, AI and green 6G network-enabled designs have the potential to revolutionize the transportation sector. For real-time data analysis and transmission between cars and infrastructure, these technologies rely on complex algorithms, sensors, and communication networks. Initiatives like the VENTURER project have illustrated the potential of AI and green 6G network-enabled architectures for autonomous connected vehicles, despite the fact that there are still a number of obstacles to be solved, including technological, social, and legal difficulties. We may anticipate major advancements in transportation efficiency, safety, and sustainability in the years to come with ongoing investment in research and development. The potential to change how we live and work in the future is represented by AI and green 6G network-enabled designs, which represent a significant shift in the way we think about mobility. We can create a more sustainable and equitable transportation system that benefits everyone by utilizing the power of these technologies.

References

[1] Li, Chen, Weisi Guo, Schyler Chengyao Sun, Saba Al-Rubaye, and Antonios Tsourdos. "Trustworthy deep learning in 6G-enabled mass autonomy: From concept to quality-of-trust key performance indicators." IEEE Vehicular Technology Magazine 15, no. 4 (2020): 112-121.

[2] Shen, Xiao, Wenrui Liao, and Qi Yin. "A Novel Wireless Resource Management for the 6G-Enabled High-Density Internet of Things." IEEE Wireless Communications 29, no. 1 (2022): 32-39.

[3] Khan, Adnan Shahid, Muhammad Ali Sattar, Kashif Nisar, Ag Asri Ag Ibrahim, Noralifah Binti Annuar, Johari bin Abdullah, and Shuaib Karim Memon. "A Survey on 6G Enabled Light Weight Authentication Protocol for UAVs, Security, Open Research Issues and Future Directions." Applied Sciences 13, no. 1 (2023): 277.

[4] Tang, Fengxiao, Bomin Mao, Nei Kato, and Guan Gui. "Comprehensive survey on machine learning in vehicular network: technology, applications and challenges." IEEE Communications Surveys & Tutorials 23, no. 3 (2021): 2027-2057.

[5] Zhu, Weilong, Chunsheng Zhu, and Zhiyun Lin. "A Prediction-based Route Guidance Method Toward Intelligent and Green Transportation System." IEEE Internet of Things Journal (2023).

[6] Li, Nan, Xiaofei Xu, Qi Sun, Jie Wu, Qiao Zhang, Gangyi Chi, and Nurit Sprecher. "Transforming the 5G RAN With Innovation: The Confluence of Cloud Native and Intelligence." IEEE Access 11 (2023): 4443-4454.

[7] Khasawneh, Ahmad M., Mamoun Abu Helou, Aanchal Khatri, Geetika Aggarwal, Omprakash Kaiwartya, Maryam Altalhi, Waheeb Abu-Ulbeh, and Rabah AlShboul. "Service-centric heterogeneous vehicular network modeling for connected traffic environments." Sensors 22, no. 3 (2022): 1247.

[8] Almutairi, Mubarak S. "Deep learning-based solutions for 5G network and 5G-enabled Internet of vehicles: advances, meta-data analysis, and future direction." Mathematical Problems in Engineering 2022 (2022): 1-27.

[9] Zhu, Weilong, Chunsheng Zhu, and Zhiyun Lin. "A Prediction-based Route Guidance Method Toward Intelligent and Green Transportation System." IEEE Internet of Things Journal (2023).

[10] Priya Kohli, Sachin Sharma, and Priya Matta. "Secured Authentication Schemes of 6G Driven Vehicular Communication Network in Industry 5.0 Internet-of-Everything (IoE) Applications: Challenges and Opportunities." In 2022 IEEE 2nd International Conference on Mobile Networks and Wireless Communications (ICMNWC), pp. 1-5. IEEE, 2022.

[11] Piyush Agarwal, Sachin Sharma, and Priya Matta. "Big Data Technologies in UAV's Traffic Management System: Importance, Benefits, Challenges and Applications." Autonomous Vehicles: Using Machine Intelligence (2023): 181.

[12] Amit Juyal, Sachin Sharma, and Priya Matta. "Multiclass Objects Localization Using Deep Learning Technique in Autonomous Vehicle." In 2022 6th International Conference on Computation System and Information Technology for Sustainable Solutions (CSITSS), pp. 1-6. IEEE, 2022.

[13] Vartika Agarwal, and Sachin Sharma. "Q Learning Algorithm for Network Resource Management in Vehicular Communication Network." Autonomous Vehicles Volume 2: Smart Vehicles (2022): 259-274.

[14] Amit Juyal, Sachin Sharma, and Priya Matta. "Anomalous Activity Detection Using Deep Learning Techniques in Autonomous Vehicles." Autonomous Vehicles: Using Machine Intelligence (2023): 1

[15] Ranu Tyagi, Sachin Sharma, and Seshadri Mohan. "Blockchain Enabled Intelligent Digital Forensics System for Autonomous Connected Vehicles." In 2022 International Conference on Communication, Computing and Internet of Things (IC3IoT), pp. 1-6. IEEE, 2022.

[16] Amit Juyal, Sachin Sharma, and Priya Matta. "Traffic Sign Detection using Deep Learning Techniques in Autonomous Vehicles." In 2021 International Conference on Innovative Computing, Intelligent

Communication and Smart Electrical Systems (ICSES), pp. 1-7. IEEE, 2021.

[17] Vartika Agarwal, and Sachin Sharma. "Forecasting-based Authentication Schemes for Network Resource Management in Vehicular Communication Network." In Intelligent Cyber-Physical Systems Security for Industry 4.0, pp. 239-256. Chapman and Hall/CRC.

[18] Amit Juyal, Sachin Sharma, and Priya Matta. "Deep Learning Methods for Object Detection in Autonomous Vehicles." In 2021 5th International Conference on Trends in Electronics and Informatics (ICOEI), pp. 751-755. IEEE, 2021.

[19] Vartika Agarwal, Sachin Sharma, "EMVD: Efficient Multitype Vehicle Detection Algorithm Using Deep Learning Approach in Vehicular Communication Network for Radio Resource Management." International Journal of Image, Graphics and Signal Processing, 2022, Volume 14, Issue 2, Pages 25-37.

[20] Vartika Agarwal, and Sachin Sharma. "P2PCPM: Point to Point Critical Path Monitoring Based Denial of Service Attack Detection for Vehicular Communication Network Resource Management." International Journal of Computing and Digital Systems 12, no. 1 (2022): 1305-1314.

Biographies

 **Sachin Sharma** is currently an associate dean, International Affairs and Professor, with the Department of Computer Science and Engineering at Graphic Era Deemed to be University, Dehradun, UK, India. He is also a co-founder and Chief Technology Officer (CTO) of IntelliNexus LLC, Arkansas, a US-based company. He also worked as a Senior Systems Engineer at Belkin International, Inc., Irvine, California, USA for two years. He received his Philosophy of Doctorate (Ph.D.) degree in engineering science and systems specialization in systems engineering from University of Arkansas at Little Rock, USA with 4.0 out of 4.0 GPA and M.S. degree in systems engineering from University of Arkansas at Little Rock with 4.0 out of 4.0 GPA. He received his B.Tech. degree from SRM University, Chennai including two years at University of Arkansas at Little Rock, USA as an International Exchange Student. His research interests include wireless communication networks, IoT, vehicular ad hoc networking, and network security.

Ranu Tyagi received her Masters of Computer Applications degree from Graphic Era Deemed to be University, Dehradun, Uttarakhand, India. Her research interests include wireless communication networks, IoT, AI, and network security.

Seshadri Mohan is currently a professor with the Systems Engineering Department at the University of Arkansas at Little Rock. Prior to his current position, he served as the Chief Technology Officer with Telsima, Santa Clara, California; as Chief Technology Officer with Comverse, Wakefield, Massachusetts; as a Senior Research Scientist, with Telcordia, Morristown, NJ; and as a member of the technical staff with Bell Laboratories, Holmdel, NJ. Besides his industry positions, he also served as an associate professor at Clarkson University and as an assistant professor at Wayne State University. Dr. Mohan has authored/coauthored over 85 publications in the form of books, patents, and papers in referred journals and conference proceedings. He has coauthored the textbook "Source and Channel Coding: An Algorithmic Approach." He holds 14 US and international patents in the area of wireless location management and authentication strategies as well as in the area of enhanced services for wireless. He is the recipient of the SAIC Publication Prize for Information and Communications Technology. He has served or is serving on the editorial boards of IEEE Personal Communications, IEEE Surveys, and IEEE Communications Magazine and has chaired sessions in many international conferences and workshops. He has also served as a guest editor for several special issues of IEEE Network, IEEE Communications Magazine, and ACM MONET. He was nominated for 2006 GWEC's Global Wireless Educator of the Year Award and selected the runner up, the 2007 ASEE Midwest Section Dean's Award for Outstanding Service. He was awarded 2010 IEEE Region 5 Outstanding Engineering Educator Award in which he was selected the runner up. Dr. Mohan holds a Ph.D. degree in electrical and computer engineering from McMaster University, Canada, the Master's degree in electrical engineering from the Indian Institute of Technology, Kanpur, India, and the Bachelor's degree in Electronics and Telecommunications from the University of Madras, India.

10

Latest Advances on Deterministic Wired/Wireless Industrial Networks

Rute C. Sofia

Research Institute Fortiss,
Email: sofia@fortiss.org

Abstract

This chapter investigates the most recent advances in deterministic OFDMA-based Wi-Fi for industrial applications. It concentrates on techniques that provide deterministic support in industrial networks where Wi-Fi 6 is used as an additional technology to a deterministic wired core. It examines scenarios and traffic patterns that are expected to be supported and proposes enhancements to further support deterministic guarantees in a way that is compatible with the existing deterministic support in the wired core. Furthermore, the chapter introduces the network simulator v3 DetNetWiFi framework, which was developed based on a realistic deterministic wired/wireless demonstrator, and which can be used to validate and experiment on a wired/wireless infrastructure that incorporates novel mechanisms for time synchronization, time-aware scheduling, and multi-AP co-OFDMA resource coordination.

Keywords: TSN, IIoT, Wireless, OFDMA

10.1 Introduction

IEEE 802.11ax, also known as Wi-Fi 6 [3], is being applied in the context of industrial environments as a complementary technology to a 5G network and also to an Ethernet/time sensitive networking (TSN) core, to support deterministic guarantees such as bounded latency and jitter, zero packet loss, required by critical and time sensitive industrial applications. The reason for the interest in considering Wi-Fi in industrial environments with the inherent flexibility of the shared wireless medium, which facilitates, for instance, the process of interconnecting a large number of Industrial IoT (IIoT) devices throughout the edge-cloud continuum, reducing costs, and facilitating the interconnection of mobile devices, e.g., automated mobile robots (AMR) or automated guided vehicles (AGV).

Wi-Fi 6 is based on orthogonal frequency-division multiple access (OFDMA). OFDMA brings in several benefits, such as the possibility for traffic isolation and simultaneous transmission to and from wireless clients based on sub-frequencies; larger channel capacity; reduced energy consumption. The possibility to consider multi-frequency ranges (2.4 GHz; 5 GHz; 6 GHz) increases the overall available spectrum for multi-user environments, being beneficial to support bandwidth-intensive applications such as high-definition video, virtual reality (VR), or augmented reality (AR), along with time-sensitive, critical applications.

In terms of energy consumption, Wi-Fi 6 integrates the target wake time (TWT) mechanism [34] to allow controllers (access points, AP) to best schedule transmissions with their wireless clients, and therefore, clients can spare energy, as we shall discuss in the next sections.

These features, among others, make Wi-Fi 6 a relevant technology to support communication in industrial environments. In such environments, IIoT applications often require strict (deterministic) guarantees in the form of zero packet loss, bounded latency, bounded jitter, and guaranteed bandwidth. In wired networks and cellular networks, these guarantees are provided today via the IEEE Ethernet standard extensions for time sensitive networking (TSN) [16].

Hence, in industrial environments, a key requirement for the successful integration of wireless technologies concerns its capability to provide low latency, low jitter, zero packet loss in a way that is compatible with the wired TSN-based regions of industrial infrastructures.

This chapter focuses on the development of deterministic wireless approaches that can benefit current and future industrial networks, having in mind backward compatibility with Ethernet/TSN standards.

The chapter contributions are:

- Provides background on the needs and enablers to achieve determinism in wireless as complementary technology to Ethernet and 5G, in a way that is backward compatible with Ethernet/TSN.

- Debates on which traffic profiles to consider wired/wireless infrastructures, to allow for a smooth integration of critical traffic and time-sensitive applications.
- Explains the features of Wi-Fi 6 that are relevant in the context of supporting deterministic behavior, in a way compatible with Ethernet/TSN.
- Discusses support of tight time synchronization in wireless based on protocols that are better suited to the wireless medium.
- Discusses how to support time-aware scheduling in wireless in a way that is compatible with TSN.
- Discusses multi-AP co-OFDMA coordinated transmission approaches, to support mitigation of interference across overlapping basic service sets (OBSS).
- Presents a novel ns-3 simulation framework that integrates the proposed mechanisms for time synchronization, time-aware scheduling, and multi-AP coordination.

The chapter is organized as follows. Section 10.2 provides examples of industrial use cases that rely on Wi-Fi 6 as a complementary solution to an existing Ethernet/TSN core, and debates on traffic requirements based on prior work that has exhaustively analyzed industrial applications that require wireless support, and debates on traffic type profiles and their quality of service (QoS). Section 10.3 debates the wireless mechanisms available in Wi-Fi 6 (IEEE 802.11ax) and Wi-Fi 7 (IEEE 802.11be) and is relevant for integrating deterministic behavior in Wi-Fi, in a way that aligns with the deterministic guarantees supported by TSN/Ethernet environments. Section 10.4 is dedicated to a debate and comparison on wireless time synchronization mechanisms that can be used to extend local time synchronization with tight accuracy and precision in wireless, while section 10.5 explains how the IEEE 802.11ax RU notion works. Section 10.6 debates on scheduling approaches that can assist in the integration of time-aware scheduling (TAS) in wireless. Section 10.7 discusses Wi-Fi 6 mechanisms to support ultra-reliability, while Section 10.8 presents the novel network simulator v3 (ns-3) DetNetWiFi framework, which provides a simulation environment to experiment with industrial wired/wireless networks, based on the mechanisms discussed in this chapter. The chapter ends in Section 10.9, where the directions of research are also debated.

10.2 Traffic Profiles and Converged Industrial Networks Use-Cases

Figure 10.1 represents a generic factory environment that has a networking infrastructure composed of three OBSSs, namely, an automation plant (shop floor), an office area, and a warehouse. On the shop floor and in the warehouse, two APs provide wireless connectivity interconnected to a TSN wired domain. In this specific environment, the AP serving the office area (AP3) is interconnected with the Internet and creates interference which may jeopardize critical traffic being supported by the other APs. A different possible configuration for such an industrial environment is provided in Figure 10.2, the difference being the fact that the automation shop floor and warehouse areas are interconnected with two different wired TSN regions.

In terms of traffic profiles, the **shop floor** (yellow) integrates a production line, workers supporting the machines, and a camera for operation monitoring. Furthermore, wireless sensors may be attached to robots to provide additional feedback to workers. A worker located at a wireless station (W3) performs monitoring and may send control traffic to robots. W3 experiences interference due to the overlapping signals controlled by the different APs. Information is periodically collected (from sensors and robots, for instance) to perform remote monitoring and to assist remote maintenance. In the represented shop floor, all actors are static, so Wi-Fi is employed to provide flexibility and reduce costs (e.g., installation of sensors or cameras). Human

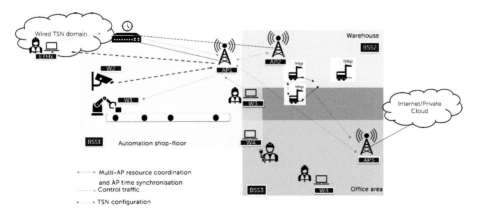

Figure 10.1 An example of a flexible factory industrial infrastructure with overlapping BSSs served by a single wired TSN region.

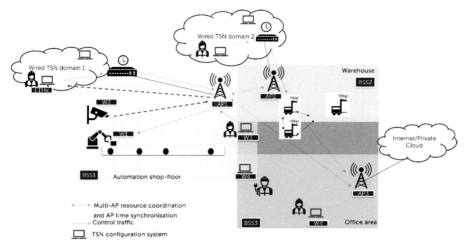

Figure 10.2 An example of a flexible factory industrial infrastructure with overlapping BSSs served by two wired TSN regions.

operators may move around the conveyor belt and robots, e.g., to inspect the robot to object positioning (UL video transmission). While image detection via camera is directed toward specific points, it is important to ensure that there is low latency in video transmission, to ensure an adequate emergency of the respective maintenance systems. Emergency traffic (e.g., alarms) can be triggered uplink (from the machines to an AP) by the conveyor belt or machine safety system; also remotely and dowlink by the worker at station W3; or remotely by a worker on another BSS or even on the wired region (TSN wired domain). Furthermore, to ensure safety, commands triggered by an emergency stop switch need to be transmitted immediately and reliably, both uplink and downlink. The shop floor is served by multiple APs interconnected to a single TSN domain. APs serving other areas, for example AP3, may create interference that can jeopardize critical traffic downlink. Between AP1 and AP2, the exchange of resources is expected to exhibit low variability, the key aspects are i) to ensure that critical traffic is adequately prioritized on both uplink and downlink directions; ii) to meet the low latency requirements of applications. Tight time synchronization is required in this case, for all APs and wireless clients. This aspect can be worked together with a multi-AP resource sharing approach. The communication in the shop floor is based on wireless infrastructure mode. Examples of traffic streams occurring in this context could be, for instance, some information being sent from W3 or from

ETH3x to the robot (downlink), characterized as control and management data – e.g., motion planning. The robot, on the other hand, may also send control and management data feedback to W3 or to ETHx, and may also send status data (monitoring) and alarms. Video may also be transmitted from the cameras on the shop floor to W3 or to ETHx.

In the **warehouse** (blue), multiple AGVs are present with autonomous driving capabilities, e.g., location and path control. Specific navigation plans and tasks to carry out are today programmed via either specific wired interfaces or via Wi-Fi based on a centralized fleet management system. Assuming environments with large fleets, there is the need for dynamic adaptation and eventual customization, which includes regular software updates and individual AGV configuration, among others. Each AGV is expected to receive commands such as "move from position x to y" (fleet control) sent from an operator in the wired TSN region or, for example, in office. AGVs may move to overlap areas (for example, between BSS2 and BSS3), thus affecting the transmission of critical traffic. Time synchronization of the end points is particularly relevant for the case of time-triggered traffic being sent downlink. Uplink, AGVs may send alerts or answers to requests, which may have to be handled with higher priority than *best effort (BE)* traffic. AGVs may also receive critical safety control data (e.g., stop to prevent a collision). The key challenge in this usage scene relates with ensuring a tight synchronization of all end-points to ensure an adequate request-response behavior. In this context, an adequate multi-AP coordination of shared resources may assist in mitigation interference.

The **office** is served by AP3, which is interconnected to the Internet and, therefore, is not directly connected or controlled via a wired TSN region. The AP is part of the AP coordinated set, i.e., it participates in the sharing of resources with the other APs, based on a selected multi-AP coordination framework. The key challenge to address in this representative example relate to the impact that an AP that cannot be directly controlled may have in terms of interference towards WLAN that serve wireless clients transmitting or receiving critical traffic, and how to mitigate such impact.

As described in the aforementioned examples, industrial networks need to support the coexistence of different applications within the same infrastructure and have the means to transport different types of traffic with appropriate QoS requirements, also ensuring backward compatibility with the requirements supported across the TSN core regions.

Defining adequate traffic profiles, their classes and priorities, as well as the respective communication requirements for each type profile, is essential

Table 10.1 Proposed traffic profiles for converged networks presented by order of priority, from highest to lowest.

Profile	PCP	Delivery guarantees	Period	Latency	Jitter	Payload (bytes)	Message type	Criticality level
Isochronous control	6	Deadline (synchronized bounded latency), zero congestion loss	<2 ms	<50% of the period	least possible	30−100	Fixed	High
Cyclic control	5	Bounded latency, zero congestion loss	<2−20 ms	<90% of the period	<1 μs	50−1000	Fixed	High
Network control	7	Bounded latency, bandwidth	50 ms-1 s	100 ms	n.a.	50−500	Variable	High
Legacy control	4	Bandwidth	>= 1 ms	< period	n.a.	50−1500	Variable	Medium
Event-based control and alarms	3	Bounded latency	Acyclic	10−50 ms (up to 2 s for alarms)	n.a.	100-1500	Variable	Medium-High
Configuration and diagnostics	2	Bandwidth, bounded latency	Acyclic	100 ms	n.a.	500−1500	Variable	Medium
Audio/video	1	Bandwidth, bounded latency	frame /sample rate	<40 ms	n.a.	1000−1500	Variable	Low
Best effort	0	None	Acyclic	n.a.	n.a.	30−1500	Variable	Low

to assist in developing the support for deterministic behavior in wireless. To provide such a definition, we have exhaustively analyzed related work in the field, namely, related industrial initiatives such as the Field Level Communications (FLC) Initiative of the OPC Foundation [36], the Industrial Internet Consortium (IIC) [6], the IEEE Nendica working group [24], the 5G-ACIA [4], or the IETF Reliable and Available Wireless (RAW) work group[1] provide characterizations of traffic in industrial environments. Furthermore, an analysis of industrial wireless applications and communication requirements has been proposed to the IETF RAW group [39].

Based on the analysis of related work, we propose the following traffic profiles, summarized in Table 10.1 along with a proposed Priority Code Point (PCP) assignment, deterministic guarantees that each profile requires, and specific communication characteristics, to provide alignment with existing TSN traffic profiles:

- **Network control**, corresponding to control plane traffic critical to the management of the network and its services. This includes time synchronization (e.g., PTP), loop breaking, and network redundancy (e.g., Multiple Spanning Tree Protocol (MSTP)), and topology discovery (e.g.,

---

[1]https://datatracker.ietf.org/wg/raw/about/

Link Layer Discovery Protocol (LLDP)). Protocol messages in this profile are expected to have a link-local scope.

- **Isochronous control**, where applications exchange synchronized data with the network in a periodic fashion. Examples include motion control, fast I/Os, or tight control loops. Messages need a guaranteed delivery time (a common deadline within the period) and must not interfere with other traffic (to achieve zero congestion loss).

- **Cyclic control**, where applications communicate periodically (cyclic) but are not (necessarily) synchronized. Examples are devices that sample inputs or set outputs periodically, where sampling and communication cycles may also differ. Cyclic control traffic may be subject to predictable interruptions from isochronous traffic, but requires a bounded latency and zero congestion loss.

- **Legacy control**. Converged TSN infrastructures will also include devices that are not TSN capable but still perform cyclic control tasks. Examples are I/Os or control loop applications with lower QoS requirements, for instance using Modbus/TCP, or devices that relied on exclusive network usage and over-provisioning.

- **Event-based control and alarms** are exchanged when specific conditions in a system change and emit asynchronous messages, which can involve a single message, a set of messages or bursts. An example of this traffic is OPC UA Alarms and Conditions. Within a specific period, the network must be able to handle the traffic up to a certain number of messages without loss, corresponding to a bandwidth guarantee requirement.

- **Configuration and diagnostics**. Also defined in some sources as "Excellent effort," this profile is intended to be used for the exchange of real-time data including device and network management (e.g., Simple Network Management Protocol (SNMP), Network Configuration Protocol (NETCONF)) and diagnostics. This data is traditionally sent using TCP-based protocols that contain lost message recovery capabilities.

- **Audio/video** corresponds to the transmission of audio and/or video streams in industrial environments. This may include automated visual inspection, remote maintenance, or safety camera systems. With latency requirements, but lower than cyclic control, this profile mainly requires bandwidth guarantees and traffic shaping to avoid bursts (see credit-based shaper (CBS)). While video is becoming more important for industrial applications (e.g., remote maintenance, visual inspection), this profile is still not considered for the IEC/IEEE 60802 industrial automation profile for TSN.

- **Best effort**, standing for traffic with no timing or delivery guarantees, usually IT traffic in converged networks.

10.3 Key Mechanisms for Wireless Time Sensitive Networking

IEEE 802.11ax relies on OFDMA, 1024 quadrature amplitude modulation (QAM), target wake time (TWT), BSS coloring, among other aspects that allow for higher throughput and a higher degree of fairness in multi-user environments [14]. For better traffic isolation and to support multi-user environments, Wi-Fi 6 introduces the concept of resource unit (RU) representing sub-channels that can be reserved by an AP to serve a specific wireless client. Furthermore, relying on 1024 QAM allows wireless devices to reach a theoretically higher throughput with higher data rates.

Additional mechanisms can assist in supporting a deterministic behavior in Wi-Fi 6 as a complementary technology to Ethernet/TSN, as will be discussed in this section. To assist the discussion, the section starts by introducing in a summarized way the key aspects of TSN.

TSN standards are defined in the realm of the IEEE 802.1 TSN Task Group. TSN relies on time scheduling to provide low bounded guarantees in terms of latency and jitter, for reliable packet delivery assuming deterministic real-time applications, treating non-deterministic traffic as BE traffic.

The support of TSN standards can be grouped into four main categories of aspects that are relevant to support determinism in Ethernet and its derivatives:

- **Time synchronization** based on the 802.1AS profile for the Precision Time Protocol (PTP) [1, 19], thus ensuring that a single local reference clock is distributed across network in a master/slave basis, and claiming an accuracy in the range of 100 ns to 50 µs.
- **Traffic shaping and scheduling** based on competing priorities (IEEE 802.1 Q); frame preemption and fragmentation mechanisms (IEEE 802.1 Qbu); TDMA scheme (IEEE 802.1Qbv) to split communication opportunities into repeating and fixed-length time cycles; a time-aware shaper then defines the time used by time-sensitive frames (thus reducing the potential impact of other types of traffic).

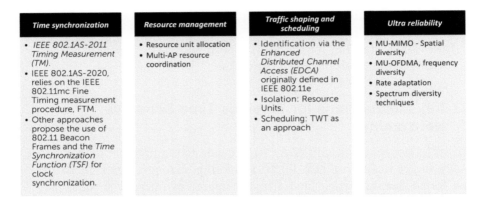

Time synchronization	Resource management	Traffic shaping and scheduling	Ultra reliability
• *IEEE 802.1AS-2011 Timing Measurement (TM).* • IEEE 802.1AS-2020, relies on the IEEE 802.11mc Fine Timing measurement procedure, FTM. • Other approaches propose the use of 802.11 Beacon Frames and the *Time Synchronization Function (TSF)* for clock synchronization.	• Resource unit allocation • Multi-AP resource coordination	• Identification via the *Enhanced Distributed Channel Access (EDCA)* originally defined in IEEE 802.11e • Isolation: Resource Units. • Scheduling: TWT as an approach	• MU-MIMO - Spatial diversity • MU-OFDMA, frequency diversity • Rate adaptation • Spectrum diversity techniques

Figure 10.3 Categories of Wi-Fi 6 mechanisms relevant to support determinism across industrial wired/wireless environments.

- **Ultra reliability** provided by IEEE 802.1CB, which requires multiple path availability (between senders and receivers). Currently, duplicate copies of each frame are sent over disjoint paths. Moreover, IEEE 802.1Qca provides path control and reservation, and 802.1Qci provides per-stream filtering and policy, thus helping to not exceed bandwidth guarantees and to reduce packet loss.
- **Resource management**, in the form of network policies to manage available resources (e.g., bandwidth, scheduling patterns) defined by IEEE 802.1Qcc, considering centralized and decentralized approaches.

Wi-Fi 6 integrates mechanisms that can assist in supporting deterministic behavior in a way that extends current TSN standards, while at the same time, better using the broadcast wireless medium. These mechanisms, summarized in Figure 10.3, are further debated in the following subsections.

10.4 Time Synchronization

The following aspects are currently the main time synchronization directions considered for Wi-Fi 6 and on the IEEE 802.11be study group, Wi-Fi 7 in a way that is compatible with wired TSN time synchronization:

- IEEE 802.1AS-2011 can already be operated over IEEE 802.11 by means of the timing measurement, via the timing measurement (TM) procedure first defined in IEEE 802.11v [31]. TM provides the means to propagate time between master and slaves. Slaves can compute the clock offset and adjust their own time accordingly.

- IEEE 802.1AS-2020 considers a variant of the Fine Timing Measurement (FTM) protocol, originally defined in IEEE 802.11mc for relative localization support [40], which claims to support 0.1 ns of timestamp resolution, against 10 ns provided by TM.
- Other approaches propose the use of 802.11 beacon frames and the time synchronization function (TSF) for clock synchronization.

TSN relies on PTP to synchronize devices based on a local area network reference clock. PTP is a point-to-point UDP-based protocol, with a claimed accuracy between 100 ns to 50 μs [19].

Related work has focused on the use of PTP in wireless networks (e.g., wireless sensor networks) [23, 30]. There have also been attempts to extend PTP in a way that is better suited to the nature of wireless shared medium, as proposed by Garg et al. with "wireless PTP," where the main aspect worked on is the reduction in overhead signaling for multihop environments [17].

In summary, it is feasible to consider PTP over wireless; however, the asynchronous delay impact needs to be accounted for in the PTP "Slave" and "Master" roles. This would also require adaptations in terms of offset computation. Furthermore, in wireless multi-user environments, PTP extensions should be considered to support multi-user synchronization.

On the other hand, wireless integrates its own time synchronization mechanisms, which can also be considered to support critical and time-sensitive applications, and which are explained next.

10.4.1 TM- and TA-based synchronization

The TM wireless mechanism has been available since IEEE 802.11v and is a MAC layer synchronization mechanism that allows 802.11 devices to synchronize to a grandmaster clock on an AP.

The TM timer has a resolution of 10 ns. Usually, the timer is based on a built-in vendor-specific timer, which can carry TAI or UTC time, and is compliant with IEEE 802.1AS-2011. The accuracy of TM, based on hardware timestamping, is claimed to be in the range of a few nanoseconds [3]. TM relies on the MAC layer management entity (MLME) interface, thus relying on common frames and also making use of measurement frames all within MLME.

10.4.2 TSF-based synchronization

IEEE 802.11 supports TSF being all Wi-Fi-enabled devices equipped with a TSF timer, i.e., a 64-bit hardware counter. In the wireless infrastructure mode, an AP periodically synchronizes clients through dedicated management frames (beacons and probe responses for Wi-Fi 5). Beacon frames are periodically sent (by default, 100 ms) and contain a 64-bit timestamp that holds the current value of the TSF timer of the AP. This timestamp corresponds to the TSF timer value when the first bit of the timestamp field of the beacon frames are transmitted via the OSI PHY layer, plus the delay of sending the data stream to the physical channel. Wireless clients in the range of the AP get the beacons, extract the timestamp, and compare it with their own TSF timer value. The client then subtracts the difference from its TSF counter, thus being able to detect a variation in offset and if required synchronizing with the AP. TSF timers are used only to synchronize clients and APs. Probe response frames can also be used to perform TSF timer synchronization. However, these frames are only exchanged at the instant when a client joins a BSS. From then on, periodic beacons assist synchronization. TSF timers can help synchronize client clocks with an AP with a precision below 1 ms[28]. It should be noted that TSF works adequately within 1 BSS and for infrastructure mode, assuming that interference in the area is reduced. Moreover, the TSF-based synchronization does not include any propagation delay estimation for the offset estimation. TSF therefore considers that the master clock resides in one AP.

10.4.3 Fine time measurement

The FTM protocol is currently defined for two purposes. The first is localization as defined in its original form in IEEE 802.11mc (IEEE 802.11-2016) [20], which provides support for time-of-flight positioning. Through FTM, a device (FTM initiator) can determine its distance from another device (FTM responder), within an accuracy range of meters, based on time of flight. Compared to signal strength propagation loss approaches, relying on the time-of-flight (Round Trip Time measurement) FTM is linearly dependent on the range. Figure 10.4 illustrates the basic FTM communication sequence for an FTM initiator that resides on a wireless client, STA1, and an FTM responder that resides on an AP.

It should be highlighted that FTM requires repeated measurements, as the RTT measurements are not completely accurate, depending on aspects

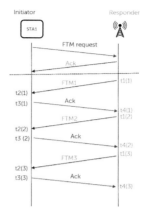

Figure 10.4   Simplified FTM communication sequence, where STA1 represents a wireless client (FTM initiator), and an AP is the FTM responder.

such as: RF interference, position of the clients toward the AP, whether the measurement is carried out in Line of Sight (LoS) or non-LoS (nLoS).

In Figure, STA1 starts by sending an FTM request to the AP. Based on the answer of the AP (Ack or no Ack), FTM will continue the process. If the AP sends an Ack acknowledging the FTM request, then the AP (responder) sends an FTM message (FTM1). This process may then be repeated (provided that the ACks are adequately received by the responder).

The FTM computation of the *RTT* for *n* FTM messages is provided in eqn 10.1.

$$RTT = \frac{1}{n} * \left(\sum_{k=1}^{n} t4(k) - \sum_{k=1}^{n} t1(k) \right) - \frac{1}{n} * \left(\sum_{k=1}^{n} t3(k) - \sum_{k=1}^{n} t2(k) \right). \quad (10.1)$$

FTM has the advantage of being capable of providing fine-grained accuracy and the disadvantage of still being defined for time measurement. In terms of synchronization on a wired/wireless TSN network, FTM is a relevant mechanism which relies on the MAC layer to support tight accuracy.

10.4.4 Comparison of wireless time synchronization mechanisms

Table 10.2 provides a summarized explanation of the features of the different available wireless synchronization mechanisms, adapted from [25, 29, 41]. Relevant to highlight is the integration of FTM into 802.1AS-Rev and its

Table 10.2 Summary of features for the different wireless time synchronization mechanisms.

Aspects	TSF	TM	TA	FTM	PTP
Resolution	1 μs	0 ns	n.a.	100 ps	60 ns
Integration	Automatic for infrastructure mode; requires coordination in multi-AP mode	TM request optional	Integrated	Initial FTM request (iFTM) mandatory	Not supported in all wireless NICs
Standard compli-ance	Integrated in 802.11	IEEE Std 802.11-2016	IEEE 802.11-2012	IEEE Std 802.11mc and being integrated into 802.1AS-Rev (Clause 12)	IEEE 802.1AS
Trigger	Beacons, 100 ms and possibility to reduce	No recom-mendations when TM should be sent	Beacon interval, can be reduced	FTM request frames, 10 ms	n.a.
Propagation delay com-putation	None	2-way, asymmetric	none	2-way, asymmetric	2-way symmetric
Accuracy	4 μs	few ns	0.5 μs	circa 1.3 μs	1.10 ns
Precision	< 1 μs	n.a.	2.5 μs	1-2 meters	3.10 ns

FTM accuracy has usually been provided in meters, as FTM has been firstly used for relative location

achievable resolution. Another relevant aspect concerns the ability to support the computation of asymmetric propagation delay, a feature that is only supported by TSF-based synchronization, TM, and FTM.

10.4.5 FTM-based time synchronization in converged networks

The prior analysis of existing wireless time synchronization approaches shows that FTM is an interesting protocol to support time synchronization in wireless due to the following aspects:

- FTM is a MAC layer mechanism and is therefore better adapted to the wireless medium.
- FTM can be used from AP to clients, or between clients, thus introducing flexibility in terms of time synchronization for multiple scenarios, including future multi-AP scenarios.
- The accuracy achieved compared to PTP and other wireless protocols reaches picoseconds, as provided in Table 1. The table provides a summarized explanation of the features of the different available

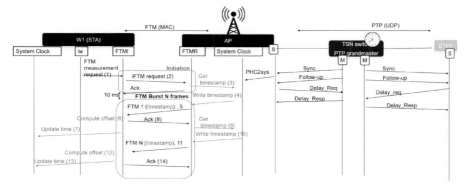

Figure 10.5 Implemented extensions to support a PTP to FTM mapping.

wireless synchronization mechanisms [1]. The integration of FTM into 802.1AS-Rev and its achievable resolution is important to highlight.

However, FTM handles the time synchronization between FTM initiators and responders within one WLAN. For the purpose of converged wired/wireless networks, there is a need to perform a mapping between the UDP-based PTP synchronization process and the FTM synchronization process. For this purpose, we have developed a proposal for the PTP-FTM mapping represented in Figure 10.5. This proposal has been implemented and validated in the fortiss IIoT Lab[2].

The figure provides a representation between one wired TSN switch holding a PTP grandmaster clock, and the wireless region represented by an AP and a wireless client (W1). In the figure, the AP WAN (Ethernet) network interface corresponds to a PTP "slave" instance. The STA corresponds to the so-called FTM initiator (FTMI), while the AP (associated with the wireless interface) has the role of FTM responder (FTMR). While PTP is UDP based, FTM is MAC layer based.

The steps for this end-to-end time synchronization process are as follows:

- W1 emits an FTM request via the Linux user-space API iw (1).
- The FTM request is passed to the driver (upper MAC layer) and sent to the AP via an IEEE80211 Action Frame (FTM Request Action Frame) (2).

---

[2]Contact the authors for the code; demonstrations are open and available on the fortiss IIoT Lab.

- On the AP, once the FTM request arrives, the AP extracts the parameters that are sent by the client W1 to negotiate the FTM process.
- If accepted, the AP FTMR (MAC layer) gets the system time (3, already synchronized via PTP) and writes the respective timestamp in the FTM response action frame, in the FTM field "variable" (4).
- Timestamps are sent on each frame passed by the AP (FTMR) to the client W1 (FTMRI).
- The FTM Response Action Frame is then passed from AP to W1 (5).
- On W1, when the FTMI receives the FTM response action frame, it receives the AP timestamp and the transmission delay via the timestamps (GP2, local timer) of the frames.
- W1 adds the AP timestamp to the transmission delay to compute the time difference (offset, 6).
- The system time is updated through FTMI (7).

A detailed explanation of the implementation and its performance evaluation is available in [18].

10.5 Resource Management

10.5.1 Resource unit allocation

In Wi-Fi 6, frequencies can be allocated considering time and frequency using the notion of RU [8, 11]. Wireless clients are assigned specific frequency chunks of the main channel. The mapping is currently performed in a static way during association negotiation. For a specific TxOp, RUs serving different wireless clients start and end at the same time. This is achieved by adding padding bits to shorter packets and requiring extra bits to be transmitted at the same power level as the data portion of the packet. For example, for a 20 MHz channel and as illustrated in Figure 10.6, if a 26-tone splitting is

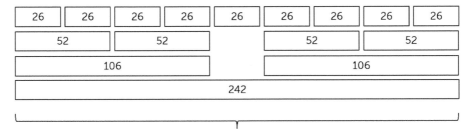

20MHz Channel

Figure 10.6   Example, Wi-Fi 6 RUs map of assignments over a 20 MHz channel.

considered, then up to nine wireless clients can be served simultaneously. If a 52-tone split is considered, then four clients can be served simultaneously. In 802.11ax, an RU can only be assigned to one wireless client. In addition, IEEE 802.11be discusses the assignment of multiple RUs to the same wireless client. Currently, the assignment map is static and is provided by specific vendors.

10.5.2 Multi-AP resource coordination

Across different BSSs, the coordination of resources is usually handled via static approaches (e.g., prior optimization of the location of APs); via IP-based coordination; or via approaches that can attempt to reduce interference, such as an automatic selection of available frequencies. Multi-AP coordination as referred to in IEEE 802.11ax and also in the IEEE 802.11be Task Group (TGbe) [21] refers to a set of features that rely on direct AP coordination to allow wireless clients across different BSSs to share a TxOp in a coordinated way, thus mitigating interference and resulting in better use of shared resources. Beyond the expected latency reduction achieved due to more efficient use of the available spectrum, multi-AP coordination can assist in protecting time-sensitive traffic across cooperating BSSs. Through a coordinated transmission approach, APs coordinate and adjust their resources and schedules in time and frequency to improve channel utilization and reduce collisions.

Coordinated transmission approaches, such as Coordinated TDMA (co-TDMA) or co-OFDMA [13], assist in a better use of resources. Through co-TDMA APs coordinate TxOP over time, i.e., they take turns transmitting to their associated wireless clients. Co-TDMA approaches require a tight time synchronization mechanism across all APs and are known to be more effective in downlink transmissions. With co-OFDMA, APs coordinate resources and schedules in time and frequency, in order to improve channel usage (reduce collisions and improve shared resource usage).

Co-OFDMA transmission schemes address the shared use of resources across time and space and are currently the ones that are being discussed in the context of IEEE TGbe to support multi-AP coordination, due to the capability to support shared resource management in time and space. While there is not yet a clear co-OFDMA mechanism to rely on, initial simulations for a probabilistic co-OFDMA approach developed by Guo et al. [22] have been presented in the IEEE [5] for a simple scenario with two OBSS serving two wireless clients. Initial simulations did not address the

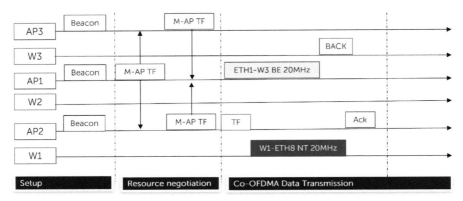

Figure 10.7   Co-OFDMA transmission example based on Figure 1.

need to integrate time-sensitive requirements across WLANs. However, the simulations showed that there are improvements for worst-case latency (3.5x reduction in maximum access time compared to EDCA), and it is the first approach that attempts to optimize time and frequency resources, showing potential in co-OFDMA approaches.

Based on the industrial scenarios presented in Section 10.1, this section proposes an architectural design for a generic co-OFDMA approach. This design is in line with the latest requirements of the IEEE TGBe specification document framework (SDF) 19/1262r23 [9] for the development of development of coordinated multi-AP transmission schemes.

We propose a generic design based on three distinct phases, namely: **setup**, **negotiation**, and **coordinated data transmission**. To assist in understanding each proposed block construction across the next sub-sections, Figure 10.7 illustrates the communication sequence for the scenario represented in Figure 10.1.

10.5.3 Setup phase

During this phase, APs discover neighboring APs via regular beacon frames. This can be done following a centralized or a decentralized approach. In the example described, we follow the decentralized design. Wireless clients sense the channel and attempt to access the wireless medium based, for instance, on hybrid coordination function (HCF)/EDCA. APs are responsible for configuring parameters such as default settings content window (CW) and EDCA access categories (ACs), in order to provide wireless clients with the best possible TxOp, while, for example, considering aspects such as traffic priority. For instance, the access to the channel may be engineered so that high-priority traffic can be transmitted first, as shall be discussed in Section 6. In this phase, in addition to regular beacon frames, APs may rely on TFs to support the exchange of control data. However, beacon frames, which are exchanged by default every 102.4 ms, can already carry information to the set of coordinating APs, along with regularly exchanged information. The beacons can transport this information based, for example, on vendor-specific information element (IE) fields. Typical examples of initial parameters that could be carried include timestamps, initial time synchronization; identifiers of individual TSN regions; channel QoS; initial RU maps, etc. For the example provided in Figure 10.7, AP1 accesses first the medium and thus sends a beacon to all other APs. AP2 and AP3 then send a response beacon which carries a multi-AP group identifier and information about active clients, and QoS other parameters that may assist APs in better assessing transmission possibilities. Response beacons may also carry information concerning clients identified as being part of in overlapping regions (e.g., W3 in Figure 1) and their assignment to a specific AP. Moreover, assuming multi-domain TSN regions, the identifiers of the specific TSN region could be carried as well.

10.5.4 Resource negotiation phase

After a set of APs has been selected or configured to handle coordinated transmission, in this phase, APs orchestrate access to the shared channel. For this, APs may exchange information to establish communication links among themselves. In this context, TGbe advocates the use of beacons or TFs to transmit the desired information. We believe that TFs are more adequate, as Wi-Fi 6 integrates a trigger-based communication approach for data transmission. The exchange of TFs allows for a faster and more natural exchange of information across the different APs.

Following Figure 7, AP1 (which in the example is the AP that would first access the medium) sends a multi-AP TF that could include a few or all of the following parameters: type of multi-AP coordination scheme (co-OFDMA); RU map; specific AP capabilities; served and active STAs; TSN scheduling parameters, e.g., TSN cycle (assuming a queuing discipline such as time-aware scheduling); time synchronization parameters, e.g., MAC of the preferred FTMR AP; channel conditions, e.g., available spectrum.

On a TF, the field to carry this information would be the "User Information" field of a TF. Currently, we expect to have each AP sending a multi-AP TF and therefore the resource negotiation phase would only be completed when all APs would have exchanged multi-AP TFs, or after a specific time window.

10.5.5 Coordinated data transmission

During data transmission, the payload (PPDU from the MAC layer perspective) is exchanged between clients and APs within a specific BSS. The AP controls communication. Therefore, to coordinate the uplink OFDMA transmissions, the AP sends a multi-user TF to all wireless clients. The parameters carried by this frame result from the negotiation phase. Specifically, this TF could indicate the number of spatial flows and (frequency and RU size) assigned to each client. It could also carry details of power control, allowing individual clients to increase or decrease their Tx power, to equalize the power the AP receives from all uplink users. Once the AP receives the frames from all clients, it sends them back a block ACK to finish the operation.

Following the example provided in Fig. 10.7, W1 sends best effort traffic to W3 via AP1, while W1 sends high-priority traffic via AP2 to W3. The selected co-OFDMA scheme will assist in resource sharing and interference mitigation. Different solutions can be applied in this context. Currently, in the IETF a probabilistic approach has been discussed for the case of two APs, where a main channel is split equally for use by the two APs based on probabilities [22]. Guo et al. mention that prior coordination of resources may further assist in the support of deterministic behavior. We have considered this approach for the purpose of experimentation; however, a probabilistic approach is not sufficient to support adequate resource sharing in OBSSs. For an implementation of the probabilistic approach on the ns-3 simulator, rf. to Section 8.

An interpretation of this probabilistic co-OFDMA approach has been further formulated by Ahn et al. again for two OBSSs, where both APs share a primary channel [7]. Their proposed formulation is relevant as it provides a first step to achieve an overall co-OFDMA proposal that considers TUA, the MU cascading sequence, and the MU EDCA to allocate TxOP to APs in a way that improves overall throughput. However, it is also important to consider latency and jitter reduction while at the same time aiming for better fairness.

AP coordination based on co-OFDMA has also been discussed in the context of multiband operation and multilink operation. For instance, Yang et al. analyze different AP coordination schemes in the context of multiband operation, and have evaluated via simulations a scheme to support multiband aggregation and multiband channel bonding [42] showing throughput improvements.

Peng et al. propose a multi-BSS multi-user full duplex scheme for handling interference mitigation across OBSS [37]. Their proposal considers a three-stage approach, adding the needs of clients in overlapping coverage areas, considering the role of a centralized controller that interacts with all APs and thus has a global view on the topology and resources. This is a static approach that may serve the purpose of the scenario represented in Figure 1, but may not suit the scenario represented in Figure 2 (multiple TSN regions).

Du et al. propose a multiple AP exponential backoff algorithm, which provides multiple APs to serve one STA simultaneously, therefore increasing resilience and system efficiency [15]. Such a mechanism may be relevant to improve system efficiency when considering clients in interference regions, where an AP may be able to provide resources to a client that is currently being served by another AP, experiencing worse conditions.

Our current research direction targets the multi-AP co-OFDMA coordination as a distributed average consensus problem. The main purpose of such a consensus approach would be to allow APs to converge by exchanging partial (BSS) information. The main objective of such a distributed consensus approach would be to compute the mean common value, i.e., the average of the initial values of measurement in a distributed way. Each iteration would comprise an exchange between the APs, updating their values through a linear combination of their own values and those of their neighbors (other APs). For an undirected fully connected graph G that considers as vertices three APs, AP1, AP2, AP3, a generic distributed consensus approach is represented by Algorithm 6 [35]. Consensus is reached when all nodes agree on the value $\mu = \frac{1}{N} * \sum_{i=1}^{N} x_i(0)$, i.e., the average of the initial values. Hence, APs can be considered as agents that carry specific state information. The initial values $x(0)$ could represent a vector of initial states, for example, a vector with information on channel QoS (e.g., RSSI, SNR); MAC QoS parameters (e.g., TSN traffic profiles supported; RU maps); number of active wireless clients served.

Algorithm 1: Distributed average consensus approach for three APs

---

Input:

      Network topology modeled with a graph $G(V, E) = (3,3)$

      Initial values for AP1, AP2, AP3 respectively,

      $x(0) : x_1(0), x_2(0), x_3(0)$

      Maximum number of iterations $k_{\max} = 2$

Data: Weight consensus matrix textitW(textitk), textitk = 0, 1, ...

1 set $k = 0$

2 $t = 0$: $AP_i, i \in \mathbb{N}$ initializes $x_i(0)$

3 **while** $k < k_{max}$ **do**

4     $x_i(k+1) = w_i i(k) * x_i(k) + \sum_{j \in N_i} w_i j(k) * x_j(k), i = 1, ..., N$

5   k=k+1

6 **end while**

      Output: x($k_{\max}$) = μ is the average of the initial value

---

10.6 Traffic Shaping and Scheduling

The following mechanisms are relevant to support traffic shaping and scheduling, considering backward compatibility approaches to TSN:

- **Identification**: EDCA, originally defined in IEEE 802.11e and integrated in IEEE 802.11ax with multi-user support (MU EDCA) provide four QoS classes. IEEE RTA provides a proposed mapping of the eight original IEEE 802.1Q priorities to the EDCA (WMM) "User policies." Currently, under discussion in TGbe is the possibility of adopting a modified priority tagging system, which would be based on the differentiated services code point (DSCP) scheme. This would be possible within the context of a WLAN only and would require the support of a specific traffic identification and classification system.

- **Preemption**: Frame preemption is a feasible alternative for traffic isolation. However, this still requires addressing fragmentation methods, methods to preserve the integrity of preemptable traffic, as shall be discussed.

- **Isolation**: Wi-Fi 6 relies on the notion of RU to provide traffic isolation, thus supporting multi-user transmissions over time and space. Isolation can also be partially supported by considering TWT.

10.6.1 Traffic identification

Wi-Fi 6 integrates MU EDCA, which provides support for legacy and Wi-Fi 6 clients to compete for the medium in fairness. When only IEEE 802.11ax is present, Wi-Fi 6 data transmission is trigger based, that is, the AP synchronizes the uplink and downlink transmissions through trigger frames (TF). IEEE802.11ax therefore provides access fairness for legacy wireless clients. Moreover, wireless clients that support MU EDCA rely on a different EDCA parameter set, the MU EDCA parameter set, which supports the possibility to consider longer contention window periods and longer backoff periods.

The operation works as follows. After being triggered by the AP to send data, a Wi-Fi 6 client may opt to rely on MU EDCA. The MU EDCA parameter set has a longer contention waiting time and a larger backoff window, and therefore has a lower priority compared to the conventional EDCA parameter set. The default values for IEEE 802.11ax MU EDCA are provided in Table 10.3.

Table 10.3   IEEE 802.11ax MU EDCA parameter set.

Parameter	Description	Value
ac-vo	Indicates AC_VO packets.	-
ac-vi	Indicates AC_VI packets.	-
ac-be	Indicates AC_BE packets.	-
ac-bk	Indicates AC_BK packets.	-
aifsn aifsn-value	Specifies the arbitration inter frame spacing number (AIFSN), which determines the channel idle time. The value is an integer that ranges from 0 to 15 and must be less than or equal to ecwmax-value.	The value is an integer that ranges from 2 to 15. ecwmin ecwmin value.
Specifies the exponent form of the minimum contention window. The ecwmin value and the ecwmax value determine the average backoff time.	Specifies the exponent form of the maximum contention window. The ecwmax ecwmax-value determine the average back-off time.	The value is an integer that ranges from 0 to 15 and must be greater than or equal to the ecwmin value.
muedcatimer muedcatimer specifies the MU EDCA timer, indicating the validity duration of MU EDCA parameters. When the timer is 0, the conventional EDCA parameters take effect.	The value is an integer that ranges from 1 to 255, in the unit of 8 TUs (1 TU 1024 μs).	

In addition to an increase in fairness in terms of channel access, MU EDCA can be considered to improve channel access for critical traffic, that is, for wireless clients whose traffic is labeled as having a higher priority. In this situation, groups of wireless clients could switch to a pre-configured MU EDCA parameter set provided by an AP. The application of MU EDCA for specific situations, e.g., transmission of critical traffic in emergency situations [10] has been tested in terms of impact for OFDM.

10.6.2 Traffic preemption

In terms of **frame preemption**, IEEE defines the Ethernet Frame Preemption based on two TSN standards, IEEE 802.3br and IEEE 802.1Qbu (Frame Preemption and Interspersing Express Traffic) [1].

The mechanisms detailed in the TSN standards guarantee a constant cycle time for critical applications and the tight control loop in industrial applications. Frame preemption mechanisms solve issues related to messages that are too long and may not fit into a TSN cycle. The purpose is to stop the transmission of long messages before the end of guard bands and continue the transmission of the rest of the frame in the TSN next cycle, after higher priority frames (express traffic) have been sent. Preempted frames are sent in multiple fragments (at least two); then, on the receiver, the fragments are assembled to recreate a copy of the original network message. Hence, frame preemption is supported via OSI MAC and PHY layer mechanisms. The PHY layer supports the interspersing express traffic (IEEE 802.3br) mechanism, while management and configuration for frame preemption are covered in the MAC layer (IEEE 802.1Qbu).

IEEE 802.3br defines a mechanism so that bridge egress ports are separated into MAC service interfaces, a preemptable MAC (pMAC) service interface and an express MAC (eMAC) service interface. Then, the MAC merge sub-layer supports interspersing express traffic with preemtable traffic. This is achieved by using a MAC merge sublayer to attach an express media access control (MAC) and a preemptable MAC to a single reconciliation sublayer (RS) service [33]. Hence, TSN traffic classes can be mapped either to the express or preemptable MAC interfaces.

Preemption implies fragmentation for preempted frames, which is handled as usual via the usual MAC layer mechanisms. Preempted frames are fragmented and reassembled in the MAC layer so that the PPDU passed to the physical layer is handled as usual. The PPDU contains the split payload and have slightly different formats, so that the first middle and last fragment of the preempted format can differ.

A similar behavior can be supported in Wi-Fi, requiring modifications to both the OSI PHY and MAC layers. As a need to frame preemption may arise at any stage of the wireless transmission process.

For instance, during channel access, different wireless clients contend for a transmission opportunity. Assuming there is a group of wireless clients marked as critical and relying on different MU EDCA sets, then this group may be provided access to the medium before other wireless clients, by playing with inter-frame spacing (IFS),contention window intervals, and backoff timers. The shorter the IFS, the higher the probability that a specific node will access the channel (a starvation of other nodes might occur). The CW size has an influence on how fast a node can try to access the channel again in case the channel was sensed busy before. It is also feasible to reduce (if required) the negotiation process before data transmission, as some frames are optional in Wi-Fi 6.

10.6.3 Traffic Isolation

Traffic isolation can be supported through RU assignment, as well as TWT. Figure 10.8 provides an example of RU assignment for the specific case of three different types of traffic: cyclic control; audio/video and best effort. Upon reception of a critical flow (e.g., a cyclic TSN flow from the wired region), the AP checks its buffer and assigns the flow to the RUs, based on the stream's destination (STA). For example, RU 1 is assigned to client STA 1, while RU 2 is assigned to STA 2 and RU 3 is assigned to STA 3. The AP checks the respective allocation and then sends a MU-RTS trigger frame with the RU map to clients. Each client responds with a CTS frame that is sent simultaneously on the assigned RU. A SIFS after the CTS, the MU-PPDUs are transmitted in parallel on the RUs that have been assigned by the MU-RTS trigger frame.

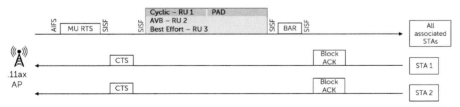

Figure 10.8 Downlink transmission, where three TSN flows are transmitted in different RUs mapped to different STAs.

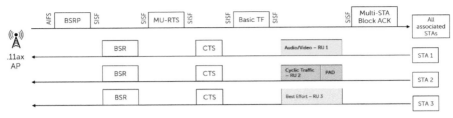

Figure 10.9   Uplink sequence diagram for traffic transmission, where traffic from different STAs is isolated per RU. In this example, different TSN streams are illustrated; however, different STAs can also simultaneously send streams belonging to the same class.

Similarly, the uplink transmission process is illustrated in Figure 10.9. The AP sends a BSRP TF, and clients in its range answer with a BSR that comprises their transmission requirements. Once the AP receives the BSRs, it handles the RU allocation with a direct mapping of RUs to clients. After the simultaneously transmitted CTSs from the clients to the AP on the respective RUs, the AP sends a basic TF to all the clients, after which the clients start sending their PPDUs simultaneously. Once that PPDU transmission is finished, the AP sends a multi-STA block ACK to the clients acknowledging the successful transmissions.

The direct mapping of RUs to clients is a simple traffic isolation mechanism, which per se cannot provide support deterministic guarantees in terms of latency and jitter. Such a direct mapping has several limitations. Firstly, it limits the number of clients that can be served simultaneously (and, as a consequence, the number of TSN flows to be served simultaneously). The increase in bandwidth and eventually other channel/band aggregation techniques can assist in increasing scalability. For instance, up to 76 STAs can be simultaneously served in a 160 MHz channel. Multiple RU assignment to one STA may further assist in improving scalability and reducing latency (e.g., using MU MIMO) [26]. Furthermore, the required TF exchange may increase bandwidth expenditure, which may be reduced by removing optional BSRP/BSR frames.

Another possibility that may better suit the need of WTSN in a way that is backward compatible with the Ethernet/TSN regions is to consider to perform RU assignment in a way that also takes into consideration the TSN traffic profile QoS requirements.

To exemplify, let us consider a scenario where one RU (RU1) is mapped to the TSN profile of network control; RU2 is assigned to the TSN profile of audio/video; RU3 is assigned to best effort traffic. Then, these three RUs could be used to transmit the three types of traffic to different STAs or to a single STA (as is under discussion in IEEE 802.11be). However, there is still a limitation as a single RU can only be mapped to a wireless client. Assigning the same RU to multiple STAs is currently not envisioned by the TGbe. It is also feasible to consider other variations, such as the assignment of RUs taking into consideration specific TSN flows (and not just traffic classes). Still, this will only suffice for a limited number of wireless clients. More dynamic environments require time-aware scheduling support, as debated next.

10.6.4 Flexible, time-aware scheduling support

The support of time-aware scheduling in a way that is aligned with the TSN time-aware scheduling (TAS) requires the integration of IEEE 802.1Qbv [2] in wireless. This requires also playing with the periods used for transmission on wireless, and finding a way on the frequency-domain to align such transmission periods with the TAS windows. This section summarizes prior work that has been introduced in [38], which relies on TWT to assist an approximation to time-aware scheduling.

The basic behavior of TWT has some similarities with TSN TAS. Both consider a cycle that is realized via control frames, and exhibit a similar approach of duty cycle. Generally, TWT is a relevant mechanism to support time-triggered scheduling in wireless in a similar sense as wired TAS [12, 24].

One possible way to support TAS via TWT is illustrated in Figure 10.10, where TWT SPs are used for cyclic (predictable) traffic types, and flexible TWT SPs are used for sporadic traffic types, e.g., event-based alarms [38]. The main concept is illustrated in Figure 10.10. This approach demands a more flexible way of handling the different TWT SPs, such that a different mapping of TSN traffic types to TWT agreement types and operational modes needs to be defined.

The main difference in the agreement type for the different TSN traffic types is that stream-based classes (isochronous and cyclic) use the individual TWT agreement type, whereas the class-based types (all others except isochronous and cyclic traffic) are mapped to broadcast TWT agreements. The two stream-based classes use the unannounced operational mode in order to optimize management overhead by reducing the number of necessary TFs.

The negotiation between the AP and the associated clients would occur in a time window which is reserved for the negotiation itself. Note that explicit (i.e., aperiodic) SPs are negotiated during each cycle depending on

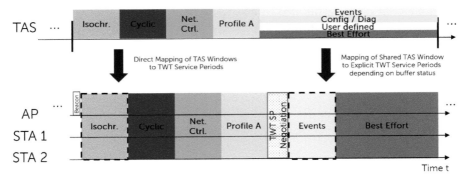

Figure 10.10 The first part of the cycle has periodically scheduled SPs, whereas the second part explicitly schedules SPs depending on the buffer status.

the buffered data. As events use the more strict approach for scheduling with dictate negotiation messages, it might also happen that, e.g., user defined, or best effort traffic may become greedy.

10.7 Ultra Reliability

Ultra reliability is handled considering rate adaptation mechanisms as well as multiplexing client transmission solutions (e.g., D-MIMO). The creation of redundant, disjoint paths is currently being considered based on the initial idea of IEEE 802.1CB, and considering additional features such as multi-AP transmission or spectrum diversity techniques. 802.11ax provides two multi-user modes: MU-MIMO, which concerns spatial diversity; MU OFDMA, which explores frequency diversity. Both modes allow for simultaneous bidirectional communication between an AP and several clients.

As has been explained before, downlink transmissions require no specific control, as they are based on a trigger mode. Uplink, depending on the type of access (scheduled or random), the process requires APs to negotiate and check clients ready to transmit, and to assign RUs, based (if requested) on the specific QoS requirements transmitted by the wireless clients.

10.8 Simulating Wired/Wireless TSN Networks: the ns-3 DetNetWiFi Framework

The ns-3 DetNetWiFi simulation framework has been developed based on a realistic wired/wireless TSN demonstrator available at Fortiss Labs and integrates advanced concepts for time synchronization. A tutorial on the framework is available online[3]. Furthermore, the code is accessible through ns-3 and the following Git repository:[4].

The framework corresponds to an ns-3 set of modules (ns3.34) currently holding the following functional blocks:

- Modeling of TSN endpoints and TSN-enabled switches interconnected to TSN WLAN regions. Multiple TSN end points interconnected to two TSN-enabled switches; a switch interconnects to a TSN WLAN region. See the TSNWiFi example. On the wired part, integrates a time-aware shaper based on the code of Krummacker et al. [27].
- TWT-based time-aware scheduling as described in [38].
- Timing model providing support for clocks per node. This block provides individual node clocks to model synchronization aspects. On the wired region, each node can therefore synchronize its own local clock to the IEEE 1588 Grandmaster clock residing on one of the switches. Each of the wireless nodes also contains a local clock [18].
- Fine time measurement protocol-based synchronization based on the proposal described in Section 2.4 [18].
- Fine-time measurement extensions to support synchronization in multi-AP environments.
- Co-OFDMA probabilistic coordinated transmission based on the work of Gao et al. [22].
- Examples with WTSN/Ethernet environments, namely, 1 TSN wired region interconnected to 1 WTSN region and also an example for 1 TSN wired region interconnecting to 2 OBSS WTSN regions.

10.8.1 Main components

Figure 10.11 provides a representation of the DetNetWiFi ns-3 framework, where Wx corresponds to a modeled wireless TSN end point; AP corresponds to a modeled AP with TSN capabilities; TSN switch corresponds to the modeling of a TSN-enabled switch, and ETHx correspond to wired TSN end points.

In the wireless region, the ns-3 DetNetWiFi framework relies on the WiFiNetDevice class to model a wireless NIC. Such a wireless NIC counts

---

[3]IEEE WPMC 2022 tutorial available via: https://git.fortiss.org/iiot_external/detnetwifi-ns3/

[4]https://git.fortiss.org/iiot_external/detnetwifi-ns3

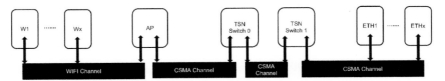

Figure 10.11   The DetNetWiFi framework components.

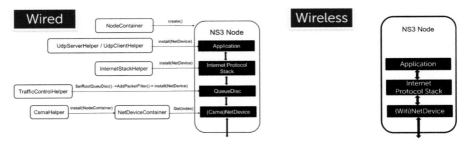

Figure 10.12   The DetNetWiFi framework, wired and wireless node objects.

with the following four layers, as illustrated in Figure 10.12: ns-3 class WiFiNetDevice:

- **PHY layer model**, ns3::WifiPhy, ns3::WifiRemoteStationManager
- **MAC lower layer**:
    o ns3::MacLow: Handles transmission aspects, e.g., RTS/CTS/DATA/ACK
    o ns3::DcfManager, ns3::DcfState, models DCF and EDCAF coordination algorithms
    o ns3::DcaTxop, ns3::EdcaTxopN, handles packet queeing, fragmentation, EDCAS, etc.
- **MAC higher layer**:
    o ns3::RegularWifiMac, which handles the AP functionality.
    o ns3::ApWifiMac, generates beacons and manages Wi-Fi associations.
    o ns3::StaWifiMac, handles association state.
- **Mobility:**[5]
    o Ns3::MobilityHelper, Helper class used to assign positions and mobility models to nodes.
    o ns3::MobilityHelper::SetPositionAllocator.
    o MobilityHelper::SetMobilityModel.

---

[5]https://www.nsnam.org/docs/models/html/mobility.html

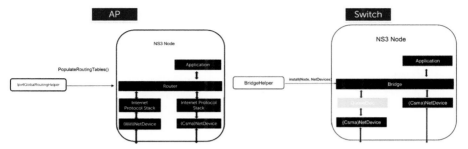

Figure 10.13   The DetNetWiFi framework, switches and AP modeling.

The modeling of the wired and wireless TSN end points is represented in Figure 10.12. TSN enabled-switches have been modeled based on existing literature, as illustrated in Figure 10.13, where the yellow box represents new code developed to support TSN TAS.

10.8.2 Timing model, clock per node

As a simulator, ns-3 is based on simulation time and therefore, by default the time is the same for all nodes. There is work in progress in this context, in ns-3[6].

We have used this proposal and improved the code, integrating the setting of clocks per node as illustrated in Figure 10.14. The adaptation we have developed considers that nodes run local clocks based on a mapping function between the node clock (local time clock) and global time (simulator time or a true global time source, e.g., NTP). The information provided in this section is also available in git[7].

10.8.3 FTM-PTP time synchronization

The proposal for FTM-based time synchronization is available in the simulation framework. The existing ns-3 FTM code[8] has been accurately developed for indoor range and localization as proposed in IEEE 802.11-2016.

---

[6]Rf. to https://www.nsnam.org/wiki/Local_Clocks,https://gitlab.com/nsnam/ns-3-dev/-/merge_requests/332/commits and https://gitlab.com/nsnam/ns-3-dev/uploads/679699e42f86 87e97589d9a0c433e3ed/ClockModule.pdf

[7]https://git.fortiss.org/iiot/detnetwifi-ns3/-/blob/main/ns-3.34/src/clock/doc/clock.rst

[8]https://github.com/tkn-tub/wifi-ftm-ns3

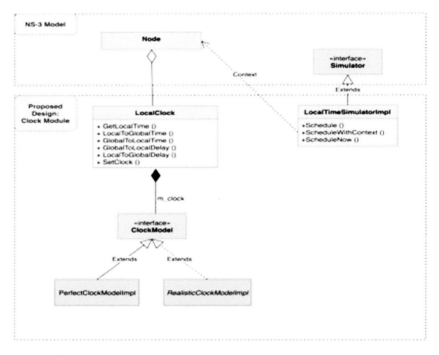

Figure 10.14   Extensions to the ns-3 clocking model, to support clocks per node.

The FTM code has been extended to allow for time synchronization, based on our proposal for PTP-FTM mapping and time synchronization, briefly introduced in Section 2.4.

10.8.4 TWT-based time aware scheduling framework

The ns-3 DetNetWiFi framework currently supports the TWT flexible scheduling approach described in Section 2.5. For this purpose, we have modeled a new API[9] so that the APs can send the TWT parameters (which represent the proposed mapping to the TSN TAS window and time slices) through a TWT action frame that can carry the different parameters (TWT wake duration; TWT wake interval; TWT wake time), as illustrated in Figure 10.15.

---

[9]https://git.fortiss.org/iiot/detnetwifi-ns3/-/blob/main/ns-3.34/scratch/TSNWifi/send-action-frame-application.cc

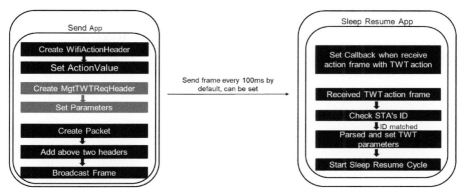

Figure 10.15   ns-3, DetNetWiFi TWT API.

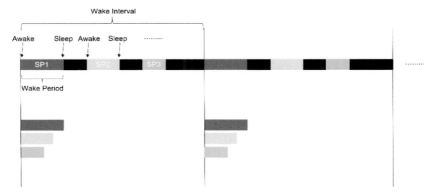

Figure 10.16   Modeling of TWT SPs on the wireless clients 1.

When an action frame carrying TWT parameters is received by a wireless client, the setting up of TWT service periods is handled by the node, via parsing. On the STA, the TWT SPs are set according to the received TWT parameters, as represented in Figure 10.16.

10.8.5 Multi-AP co-OFDMA probabilistic coordinated transmission

The co-probabilistic co-OFDMA coordination approach is also available in the DetNetWiFi simulation framework. The current implementation also considers a scenario with three APs (AP1, AP2, and AP3), where AP1 and AP2 are interconnected to the same Ethernet/TSN region. Three clients are served by the two APs: W1 is served by AP2, while W3 is served by AP1.

W2 is in an overlap region. The implementation of the probabilistic approach follows the steps detailed in the flow chart provided in Fig. 10.17.

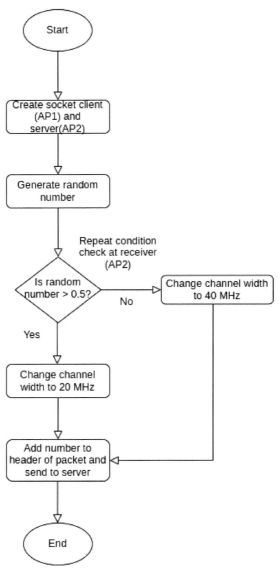

Figure 10.17   Flow chart, multi-AP co-OFDMA probabilistic approach.

After the AP contention period and once its backoff counter reaches 0, the AP randomly selects a probability p from a uniform interval and sends a MAP-TF to all its neighboring APs. The TF will trigger the channel splitting in multiples of 20 MHz, based on the total number of existing APs. The MAP-TF frame has been modeled with an ns-3 application itself is a data packet with a custom header created using the *MyHeader class*[10]. The way we have modeled the need to adapt the channel is based on the selected p and on a static threshold. The selected p is sent via the MAP-TF so that any AP knows that it needs to adjust the channel automatically. This occurs in our instantiation if p is equal to or larger than 0.5. Then, each AP changes its channel bandwidth automatically from 40 to 20 MHz. If instead p is smaller than 0.5, then the AP that receives the MAP-TF continues on the primary 40 MHz channel.

In summary, each AP selects on its own a probability p. On the basis of this p, which is transmitted by all APs via a MAP-TF to all neighboring APs, each AP can autonomously adapt its channel width.

10.9 Summary and Future Work

This work is focused on igniting discussion and proposing specific approaches to support deterministic behavior in wireless industrial environments. To better position the explanation, the chapter starts with a proposal on how to characterize traffic profiles in industrial environments, and then goes over key wireless available mechanisms that can be considered to support deterministic communication guarantees for industrial applications. In addition to the debate, the chapter explains novel proposals for time synchronization based on the FTM protocol; time-aware scheduling proposals based on TWT, in a way that is backward compatible with Ethernet/TSN standards. The chapter provides an overview of the ns-3 DetNetWiFi framework, which has been built based on a realistic wired/wireless demonstrator available on the Fortiss IIoT Labs, and which can be used by the research community to further address the analysis of deterministic requirements across wired/wireless TSN environments.

In terms of current and future research directions, the deterministic aspects of wireless integration are currently in the works. A few areas of work are as follows:

---

[10]https://www.nsnam.org/doxygen/main-packet-header_8cc_source.html

- We are currently addressing mechanisms to support a dynamic allocation of RUs based on specific TSN traffic profiles.
- We shall continue the definition of a co-OFDMA coordinated transmission mechanism based on decentralized solutions, e.g., consensus.
- The work is also being pursued to integrate WTSNs across an IP backbone (DetNet) based on a flexible, centralized DetNet control plane, considering mechanisms for flow identification and mapping; resource reservation and time synchronization, and eventual exposure of wireless features to the IP Layer.
- An integration of the current simulation environment toward a 5G core/TSN is also envisioned.

Acknowledgements

This research has been partially addressed in the context of the Fortiss projects TSNWiFi and DetNetWiFi. We thank all members of the Fortiss IIoT team for their dedication and contributions to the results achieved between 2020 and 2023.

References

[1] IEEE 802.1AS-2020 - IEEE Standard for Local and Metropolitan Area Networks - Timing and Synchronization for Time-Sensitive Applications, year=2020, organization = IEEE, note = Available online at https://standards.ieee.org/standard/802_1AS-2020.html last accessed on July 12th 2021,.

[2] IEEE 802.1Qbv - Enhancements for Scheduled Traffic. Available online at http://www.ieee802.org/1/pages/802.1bv.html last accessed on October 30th 2018.

[3] IEEE P802.11ax - IEEE Draft Standard for Information Technology – Telecommunications and Information Exchange Between Systems Local and Metropolitan Area Networks – Specific Requirements Part 11: Wireless LAN Medium Access Control (MAC) and Physical Layer (PHY) Specifications Amendment Enhancements for High Efficiency WLAN.

[4] 5G-ACIA. 5G for Connected Industries and Automation. (November), 2019.

[5] Toni Adame, Marc Carrascosa-Zamacois, and Boris Bellalta. Time-sensitive networking in ieee 802.11 be: On the way to low-latency wifi 7. *Sensors*, 21(15):4954, 2021.

[6] A. Ademaj, D. Puffer, D. Bruckner, G. Ditzel, L. Leurs, M. Stanica, P. Didier, R. Hummen, R. Blair, and T. Enzinger. Iic results white paper: Time sensitive networks for flexible manufacturing testbed characterization and mapping of converged traffic types, March 2019.

[7] Woojin Ahn. Novel multi-ap coordinated transmission scheme for 7th generation wlan 802.11 be. *Entropy*, 22(12):1426, 2020.

[8] Adnan Aijaz. High-Performance Industrial Wireless: Achieving Reliable and Deterministic Connectivity Over IEEE 802.11 WLANs. *IEEE Open Journal of the Industrial Electronics Society*, 1(April):28–37, 2020.

[9] Eduard Au. Specification Framework for TGbe - IEEE 19/1262r23. *IEEE 802.11 TGbe*, 2021.

[10] Manikanden Balakrishnan, Driss Benhaddou, Xiaojing Yuan, and Deniz Gurkan. Channel preemptive edca for emergency medium access in distributed wireless networks. *IEEE Transactions on Wireless Communications*, 8(12):5743–5748, 2009.

[11] Dmitry Bankov, Andre Didenko, Evgeny Khorov, and Andrey Lyakhov. OFDMA Uplink Scheduling in IEEE 802.11ax Networks. In *IEEE International Conference on Communications*, volume 2018-May. Institute of Electrical and Electronics Engineers Inc., jul 2018.

[12] Boris Bellalta and Katarzyna Kosek-Szott. AP-initiated multi-user transmissions in IEEE 802.11ax WLANs. *Ad Hoc Networks*, 85:145–159, mar 2019.

[13] Huanhuan Cai, Bo Li, Mao Yang, and Zhongjiang Yan. Coordinated tdma mac scheme design and performance evaluation for the next generation wlan: Ieee 802.11 be. In *Smart Grid and Internet of Things: 4th EAI International Conference, SGIoT 2020, TaiChung, Taiwan, December 5–6, 2020, Proceedings*, pages 297–306. Springer, 2021.

[14] Dave Calvacanti. IEEE 802.11: Wi-Fi 6 and Beyond. 2019.

[15] Yongliang Du, Shujie Zou, and Haixia Cui. A novel medium access control scheme with multi-ap for wifi networks. In *2020 IEEE 2nd International Conference on Civil Aviation Safety and Information Technology (ICCASIT*, pages 395–398. IEEE, 2020.

[16] Norman Finn. Introduction to Time-Sensitive Networking. *IEEE Communications Standards Magazine*, 2(2):22–28, jul 2018.

[17] Ayush Garg, Akash Yadav, Axel Sikora, and Ashok Singh Sairam. Wireless Precision Time Protocol. *IEEE Communications Letters*, 22(4):812–815, 2018.

[18] Sugandh Huthanahally Mohan and Rute Sofia. Fine time measurement based synchronisation for industrial wireless/wired networks. 09 2022.

[19] IEEE. 1588-2008 - IEEE Standard for a Precision Clock Synchronization Protocol for Networked Measurement and Control Systems.

[20] IEEE. IEEE Standard for Information technology–Telecommunications and information exchange between systems LAN and MAN–Specific requirements - Part 11: Wireless LAN Medium Access Control (MAC) and Physical Layer (PHY) Specifications, IEEE Std 802.11-2016. Technical report, 2016.

[21] IEEE. Standard for Information technology–Telecommunications and information exchange between systems Local and metropolitan area networks–Specific requirements - Part 11: Wireless LAN Medium Access Control (MAC) and Physical Layer (PHY) Specifications Amendment: Enhancements for Extremely High Throughput (EHT). accessed on 10 February 2023 2019.

[22] J. Y. Guo and Y. Li. IEEE 802.11-19/1592r0 - Simulation Results for coordinated OFDMA in multi-AP operation. *IEEE Mentor Present*, 2019.

[23] Pengyi Jia, Xianbin Wang, and Kan Zheng. Distributed Clock Synchronization Based on Intelligent Clustering in Local Area Industrial IoT Systems. *IEEE Transactions on Industrial Informatics*, 16(6):3697–3707, jun 2020.

[24] K. Meng and A. Jones and D. Cavalcanti and K. Iyer and C. Ji and K. Sakoda and A. Kishida and F. Hsu and J. Yee and L. Li and E. Khorov. IEEE 802.11 Real Time Applications TIG Report, November 2018. Available online at https://mentor.ieee.org/802.11/dcn/18/11-18-2009-06-0rta-rta-report-draft.docx last accessed on March 2021.

[25] Juha Kannisto, Timo Vanhatupa, Marko Hännikäinen, and Timo D. Hämäläinen. Precision time protocol prototype on wireless LAN. In *Lecture Notes in Computer Science (including subseries Lecture Notes in Artificial Intelligence and Lecture Notes in Bioinformatics)*, volume 3124, pages 1236–1245. Springer Verlag, 2004.

[26] Evgeny Khorov, Andrey Lyakhov, Alexander Ivanov, and Ian F. Akyildiz. Modeling of real-time multimedia streaming in Wi-Fi networks with periodic reservations. *IEEE Access*, 8:55633–55653, 2020.

[27] Dennis Krummacker and Luca Wendling. TSN Simulation: Time-Aware Shaper implemented in ns-3. In *Proceedings of the 2020 Workshop on Next Generation Networks and Applications (NGNA 2020), Kaiserslautern, Germany*, pages 16–17, 2020.

[28] Aneeq Mahmood, Reinhard Exel, and Thomas Bigler. On clock synchronization over wireless LAN using timing advertisement mechanism and TSF timers. In *ISPCS 2014 - Proceedings: 2014 International IEEE Symposium on Precision Clock Synchronization for Measurement, Control and Communication*, pages 42–46. Institute of Electrical and Electronics Engineers Inc., nov 2014.

[29] Aneeq Mahmood, Reinhard Exel, and Thilo Sauter. Performance of IEEE 802.11's Timing Advertisement Against SyncTSF for Wireless Clock Synchronization. *IEEE Transactions on Industrial Informatics*, 13(1):370–379, feb 2017.

[30] Aneeq Mahmood and Felix Ring. Clock Synchronization for Wired-Wireless Hybrid Networks using PTP. In *IEEE International Symposium on Preci-sion Clock Synchronization for Measurement, Control and Communication Proceedings*, 2012.

[31] Alexander Mildner. Time Sensitive Networking for Wireless Networks - A State of the Art Analysis. 2019.

[32] Sugandh Huthanahally Mohan and Rute C. Sofia. Fine time measurement based time synchronization for multi-ap wireless industrial environments. In *2023 19th International Conference on Wireless and Mobile Computing, Networking and Communications (WiMob)*, pages 399–404, 2023.

[33] Ahmed Nasrallah, Akhilesh S. Thyagaturu, Ziyad Alharbi, Cuixiang Wang, Xing Shao, Martin Reisslein, and Hesham ElBakoury. *Ultra-low latency (ULL) networks: The IEEE TSN and IETF DetNet standards and related 5G Ull research*, volume 21. 2019.

[34] Maddalena Nurchis and Boris Bellalta. Target wake time: Scheduled access in IEEE 802.11ax WLANs. *IEEE Wireless Communications*, 26(2):142–150, 2019.

[35] Reza Olfati-Saber and Richard M Murray. Consensus problems in networks of agents with switching topology and time-delays. *IEEE Transactions on automatic control*, 49(9):1520–1533, 2004.

[36] OPC UA Foundation. OPC UA for Field Level Communications - a Theory of Operation. Technical Report, v1, November 2020.

[37] Meiping Peng, Bo Li, and Zhongjiang Yan. Multi-bss multi-user full duplex mac protocol based on ap cooperation for the next generation wlan. *Mobile Networks and Applications*, pages 1–12, 2023.

[38] Ben Schneider, Rute C. Sofia, and Matthias Kovatsch. A proposal for time-aware scheduling in wireless industrial iot environments. In *NOMS 2022-2022 IEEE/IFIP Network Operations and Management Symposium*, page 1–6. IEEE Press, 2022.

[39] Kovatsch M. Mendes P. Sofia, R. C. Requirements for Reliable Wireless Industrial Services, 2021.

[40] Kevin Stanton, Carlos Aldana, Ashley Butterworth, Matt Mora, Michael Johas Teener, Craig Gunther, and Harman International. Addition of p802.11-MC Fine Timing Measurement (FTM) to p802.1AS-Rev:Tradeoffs and ProposalsRev 0.9. (March), 2015.

[41] Georg Von Zengen, Keno Garlichs, Yannic Schrcoder, and Lars C. Wolf. A sub-microsecond clock synchronization protocol for wireless industrial monitoring and control networks. *Proceedings of the IEEE International Conference on Industrial Technology*, pages 1266–1270, 2017.

[42] Mao Yang, Bo Li, Zhongjiang Yan, and Yuan Yan. Ap coordination and full-duplex enabled multi-band operation for the next generation wlan: Ieee 802.11 be (eht). In *2019 11th international conference on wireless communications and signal processing (WCSP)*, pages 1–7. IEEE, 2019.

11

Cyber Threat Detection in 6G Wireless Network using Ensemble Majority-Voting Classifier

Karan Sharma, Kavya Parthasarathy, Pranav M. Pawar, and Raja M.

Birla Institute of Technology and Science Pilani, UAE

Abstract

Advance tools are required for securing the 6G wireless network, like an IDS (intrusion detection system), that utilizes the current ongoing technologies as machine learning, applying which can boost up the accuracy of results, which has been discussed in this paper. This paper attempts to perform a comprehensive analysis upon application of various distinct machine learning classifiers to the AWID-(II)-2019 dataset, which contains both types of traffic flows for wireless networks, i.e., normal and abnormal, to derive which classifier has the best efficiency and accuracy in detecting intrusions. This evaluation has been conducted based on certain parameters and performance evaluation metrics, based on the construction of a confusion matrix. This chapter proposed an ensemble majority-voting classifier for intrusion detection systems, to achieve the best results based on the above-mentioned performance metrics. The accuracy of the results for the proposed ensemble majority-voting classifier is 90.45%.

Keywords: Cyber Threat Detection, Machine Learning Classifiers, Intrusion Detection System, AWID-(II)-2019 Dataset

11.1 Introduction

Due to the rapid development of computer and wireless networks including the Internet, all data and important information is readily and easily available.

255

All this information which may be communicated through wireless public networks can be intercepted and tampered with, by intruders having different intentions, thereby creating risk of permanent loss of data or compromise of data. As time passes and technology evolves, maintaining the confidentiality, integrity, availability, and privacy (CIAP) of data seems to be an important topic of concern as it is worth millions to organizations all over the world and it is imperative to continue to develop effective and efficient solutions to the problem of network security. The network security problem involving both, data security and network system security deals with preventing illegal attacks from taking place on our target system and maintaining the CIAP of data.

An important part of understanding whether a network has been compromised involves continuously scanning your network and detecting anomalies and this can be done through machine learning. Machine learning (ML), a subset of artificial intelligence (AI), allows computer systems to learn by experience the way people do, i.e., it allows a machine to learn without being programmed explicitly. As mentioned earlier, within security, ML not only helps analyze data by continuously studying the data to identify patterns to detect any anomaly or malware present in the encrypted traffic, which can also be called an intrusion into the network; but also helps in identifying insider threats and helps in protecting data stored in the cloud by flagging suspicious activity among others [1].

There are several methods to detect different anomalies within network traffic and most of the ML classifiers do so by comparing current traffic and looking for variations if present. Broadly speaking, all ML models fall within the following categories including supervised learning methods, unsupervised learning methods, reinforcement learning methods and semi-supervised learning methods. Unsupervised method of learning algorithms can make use of unlabeled datasets; understand the normal type of traffic that flows through a network and further recognize patterns alongside reporting any malicious traffic or anomalies in the normal pattern. One flaw with unlabeled datasets is the fact that there can be many cases wherein the traffic is falsely predicted as anomalous i.e., false-positive (FP) results. To reduce the chances of this happening, we use a labeled dataset for intrusion or cyber threat detection thereby using the algorithms following supervised method of learning in which the algorithm is trained so that it is able to differentiate between normal and abnormal/malicious network traffic. An advantage when it comes to using supervised machine learning models/algorithms is the fact that it can handle all attacks that are known and recognize such attacks if they seem to be present in the network [1, 2].

This chapter mainly introduces machine learning-based network security for 6G wireless communication and focuses on one of the many technologies involved in network security i.e., the intrusion detection system (IDS) and further goes on to apply ML to determine the best classifier model that can accurately classify a potential predicted threat as an actual threat. Also, this paper summarizes a plethora of different ML classifiers including decision tree, Gaussian Naïve Bayes, multinomial Naïve Bayes, random forest, logistic regression and perceptron on the chosen dataset after preprocessing and reducing it. The dataset considered here for proposed ensemble learning is AWID-(II)-2019 [1, 2]. The data has been preprocessed by performing attribute reduction i.e., reducing the number of attributes in the dataset from 155 to 20, relabeling of original class labels in the dataset by making it a binary classification problem from a multiclass classification problem i.e., all tuples were assigned to either class "normal" or class "abnormal", and lastly type conversion of all the attributes so as to make it classifier friendly. After preprocessing the dataset, various classifiers have been applied and the accuracy of each model has been individually compared to the accuracy of the proposed ensemble majority-voting classifier.

The remainder of this chapter is organized as follows: Section 11.2 discusses the related work with a comparative review of related work; Section 11.3 describes the research design methodology with dataset preprocessing, ensemble majority-voting classifier, and workflow; Section 11.4 discusses the results; Section 11.5 concludes the work.

11.2 Related Work

The section reviews the state-of-the-literature in development of IDS using machine learning techniques. Most of the research has been carried out based on the prediction of the best classifier model that detects intrusion from a set of traffic, over different datasets. The work reviews the state-of-the-art work by considering its objectives, algorithm, datasets, limitations, and performance measures used.

Suad et al. [5] introduced a model called Spark-Chi-SVM for intrusion detection (ID). This did so as they argued that big data is difficult to store and manage as they include a huge velocity of data including certain data that poses a requirement for developing new techniques to analyze them.

Moreover, since traditional intrusion detection systems (IDS) have a compli-cated analysis process and take longer time, it makes them less efficient when it comes to big data keeping the system more at risk as due to complexity, there would be a delay in generating any alert. Hence, in the model suggested, feature selection has been done using the ChiSqSelector and support vector machine (SVM) used on the Apache Spark platform for big data has been used to build the ID model. They have also drawn a comparison between the Chi-SVM classifier and the Chi-logistic regression classifier concluding that the Chi-SVM model seems to give higher performance alongside lower training time and is more efficient for big data.

Kamran et al. [6] presented their research wherein they evaluated the performance of certain ML algorithm's ability to detect various cyber threats. The ML techniques that they based the focus of their research on were decision trees, deep belief network (DBN) and SVM and the performance of these methods were evaluated by analyzing their performance in detection of intrusion, spam and malware. The datasets used for spam detection were Spambase (for email spam) and twitter database (for spam tweets) which were tested on the SVM and gave a 96.9% accuracy using Spambase and approximately 94% accuracy using the twitter database. Enron and Spambase (for email spam) tested on the decision tree classifier model gave 96% accuracy using Enron and 92.08% using Spambase. Enron and Spambase both tested on DBM gave approximately 95% accuracy. When it came to ID, the NSL-KDD dataset when tested on the SVM gave an accuracy of 89.7% and DARPA gave 95% for anomaly-based ID. Using the KDD dataset on the DBM gave 97.5% accuracy and the NSL-KDD dataset gave an accuracy of 90.4% for anomaly-based ID. When it came to malware detection, it was observed that using the malware dataset on the SVM classifier gave an accuracy of 94.37% and Enron gave an accuracy of approximately 94%. Using the malware dataset on the decision tree model gave an accuracy of 84.7% and the KDD99 dataset on the DBN gave an accuracy of 91.4%. They presented these findings alongside their methodology in their research.

In another research carried out by [7] Rishabh Das and Thomas H. Morris, the application of machine learning (ML) in cybersecurity for intrusion detection, traffic classification and email filtering is researched upon. They performed a literature survey on various data mining and ML methods and checked their application in the field of cybersecurity and thereby compared the performance of various ML methods so as to determine the best one for different security issues. To do so, they performed an evaluation of four ML algorithms on MODBUS data.

Razan et al. [8] have talked about the relevance of machine learning and deep learning approaches in prediction and detection of cyber threats. Their research is focused on the AWID-2019 dataset and have successfully been able to extract four attribute sets, and outperformed various previous related works by achieving a high accuracy score and low number of false-positive results, by application of random forest classifier and achieving a maximum accuracy score of 99.64% with supply test data and 99.9% with the 10-fold cross validation approach for the J48 and random forest classifiers.

Yuyang et al. [9], in an attempt to increase the detection ability of an intrusion detection system (IDS), have built an efficient IDS based on a vote-based ensemble classifier and meta-heuristic-based optimization algorithm for feature selection. To validate their efficiency results, they have made use of the 10-fold cross-validation (CV) method and to classify benign traffic from the rest. They have also made use of the correlation-based feature selection model and bat algorithm approach for feature selection and classifiers as C4.5 and random forest for training purposes and evaluated the performance of this over the NSL-KDD dataset, AWID-2019 dataset and the CIC-IDS-2017 dataset. Following this, various comparisons have also been made with other state-of-the-art methods for the same, in which their model seemed to outperform most of the others.

Alwi et al. [10] performed a comparative study wherein they implemented different machine learning algorithms on the NSL-KDD dataset in an attempt to see which algorithm was able to detect intrusion better. They used several classifiers including K-nearest neighbor (KNN), logistic regression, random forest, Naïve Bayes, extra-tree classifier, etc., which classified traffic as normal or malicious. Before applying the algorithms, they preprocessed the dataset and they received very good results that showed that the performance of the random forest classifier, decision tree classifier and the extra-tree classifier was more than 99% for all the different attack classes.

Tohari et al. [11] performed a study wherein the applied certain machine learning algorithms including support vector machine, K-nearest neighbor (KNN) and Naïve Bayes on different datasets including KDD Cup99, Kyoto 2006, and UNSWNB15 in an attempt to understand which algorithm can detect intrusion with the best accuracy. After preprocessing the dataset using normalization and discretization, and further applying feature selection and further implementing the algorithms on the different datasets, the results generated are compared and overall, it is seen that the support vector machine model generates the overall best accuracy of 99.9291%, true-positive rate of 99.9%, and the false-positive rate of 0%.

Table 11.1   Comparative review of state-of-the-art work.

Ref.	Objective	Dataset	Algorithm	Performance measure
[5]	Detecting intrusion on big data environment using ML classifiers	KDD 99	Spark-Chi-SVM	Training time and testing time
[6]	Evaluating the capability of ML algorithms in detecting cyber threats	Spambase, Twitter and malware database, KDD 99, Enron, NSL-KDD, DARPA	decision trees, deep belief network (DBN), SVM	Accuracy, precision, recall
[7]	Comparing the performance of various ML methods to find the best one for different security issues	MODBUS Data	Nave Bayes, random forest, OneR, J48	Area under curve, precision, recall
[8]	To utilize different methods to handle imbalanced datasets and thereby build a good intrusion detection system (IDS)	CIDDS-001	Deep neural networks, random forest (RF), voting, variational auto-encoder, stacking machine learning	Accuracy, false alarm rate, G-mean, sensitivity, combined metric
[9]	Increasing the ability of an IDS to detect any kind of malicious traffic with high accuracy and low false alarms	NSL-KDD, AWID 2019, CIC-IDS-2017	Meta-heuristic based optimization algorithm for feature selection and vote-based ensemble model	Accuracy, precision, F-measure, detection rate, attack detection rate, false alarm rate
[10]	Verifying and comparing the performance of different classifiers in classifying intrusive traffic	NSL-KDD	Support vector machine, Nave Bayes, random forest, decision tree, K-nearest neighbor, multi-layer perceptron	Accuracy
[11]	Applying data mining in order to detect anomaly in an IDS	KDD Cup 99, Kyoto 2006 and UNSWNB15	K-nearest neighbor (KNN), support vector machine, Nave Bayes	Accuracy, true-positive rate, false-positive rate

11.3 Research Methodology

The methodology that was applied to solve the problem of cyberattack detection is as shown in Figure 11.1. This discussion of the section is divided into three main parts – dataset considered, data preprocessing and ensemble classifier.

11.3.1 Dataset

The dataset being used for the research scope of this paper and for testing purposes is the Aegean WiFi intrusion dataset i.e., the AWID-(II)-2019 dataset. This dataset is publicly available over the Aegean University Website, and contains data regarding network traffic flow, which includes both normal 802.11 traffic and intrusive traffic. The chosen dataset when compared to others is oriented more towards the task of intrusion detection (in wireless networks). The traces that are present in the dataset have been taken from a real-life utilization of a protected and dedicated network. The AWID dataset

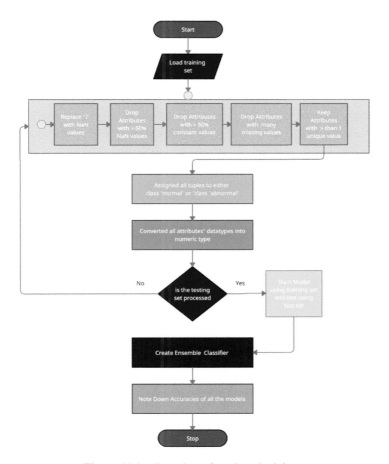

Figure 11.1   Overview of used methodology.

folder on downloading has several train-test sub-dataset files which include full and reduced versions, the one considered for the scope of this paper is the AWID-ATK-R (both train and test sets) dataset, which contained traffic eateries of type *amok, arp, authentication_request, beacon, café_latte, deauthentication, evil_twin, fragmentation, normal,* and *probe_response.* The training set has 1,795,575 rows/tuples and 155 columns/attributes and the testing set has 575,643 rows and 155 attributes.

The implemented and proposed ensemble majority-voting classifier has taken all the malicious traffic types as one class i.e., the abnormal class, and the other class corresponds to normal traffic class. For various wireless technologies, this dataset is a valuable tool when it comes to research, and hence the reason why this paper has a scope limited to the use of this dataset itself.

11.3.2 Data preprocessing

The preliminary step before applying the classifiers on the training dataset is to preprocess the dataset, which has been conducted in this work by reducing the size of the dataset, the step-by-step procedure of which is discussed below.

The first step is to import all necessary modules required to run the program smoothly. Then, the dataset was imported into a pandas data frame

```
radiotap.length                          int64
radiotap.datarate                        float64
radiotap.channel.freq                    object
radiotap.antenna                         object
wlan.fc.type_subtype                     object
wlan.fc.ds                               object
wlan.ba.bm                               object
wlan.fcs_good                            object
wlan_mgt.fixed.capabilities.ibss         object
wlan_mgt.fixed.capabilities.cfpoll.ap    object
wlan_mgt.fixed.current_ap                object
wlan_mgt.fixed.status_code               object
wlan_mgt.fixed.reason_code               object
wlan_mgt.fixed.fragment                  object
wlan_mgt.tagged.all                      object
wlan_mgt.ds.current_channel              object
wlan_mgt.tim.dtim_count                  object
wlan_mgt.tim.dtim_period                 object
wlan_mgt.rsn.version                     object
class                                    object
dtype: object
```

Figure 11.2   Data types of the remaining columns.

```
radiotap.length                          int64
radiotap.datarate                        float64
radiotap.channel.freq                    uint8
radiotap.antenna                         float64
wlan.fc.type_subtype                     uint8
wlan.fc.ds                               uint8
wlan.ba.bm                               uint8
wlan.fcs_good                            float64
wlan_mgt.fixed.capabilities.ibss         float64
wlan_mgt.fixed.capabilities.cfpoll.ap    uint8
wlan_mgt.fixed.current_ap                uint8
wlan_mgt.fixed.status_code               uint8
wlan_mgt.fixed.reason_code               uint8
wlan_mgt.fixed.fragment                  uint8
wlan_mgt.tagged.all                      float64
wlan_mgt.ds.current_channel              float64
wlan_mgt.tim.dtim_count                  float64
wlan_mgt.tim.dtim_period                 float64
wlan_mgt.rsn.version                     float64
class                                    uint8
dtype: object
```

Figure 11.3   Updated data types as numeric.

so as to perform manipulations on the data frame. On printing the data frame, it was observed that there are no column names, so firstly the column names for each attribute have been set. The total number of rows and columns in the dataset; *(1795574, 155)* ⇒ *(N(rows), N(columns))*, can be seen using the shape feature. Thereafter, all the missing or random values i.e., the rows that have attribute values as '?' are dealt with and replaced with NaN (Not a Number) values. Followed by which, the columns having more than 60% of their values as NaN values have been dropped, which reduced the number of attribute columns from 155 to 99, but doesn't change the number of tuples. Further, the columns in which all the values is some constant value are then dealt with, and which reduced the number of columns from 99 to 97. Then the columns where more than 50% of the values are constant were discarded, which further reduced the attribute count from 97 to 77. Although the decrease from 155 attributes to 77 was a good reduction in number, 77 attributes still seem like a large number to process and analyze. The next tactic used to decrease the number of columns was by checking the total number of null values in the columns and removing most of the columns that have many rows having null values. This reduced the number of columns from 77 to 22. All the attributes with NaN and missing values were replaced by the constant 0 after which, only the attributes having more than one unique value were kept, bringing the final attribute count to 20 (Figure 11.2) . All the malicious traffic labels such as *amok, arp, deauthentication, beacon*, etc., are grouped and replaced under one category or class label, i.e., *abnormal*, and the remaining as *normal*. The observations of the reduction along with the data types of the remaining columns can be seen in Figure 11.3 post converting data types to numeric types.

11.3.3 Ensemble majority-voting classifier

Figure 11.4 shows the architecture and process of building ensemble majority-voting classifier using seven different classifiers (decision tree, logistic regression, Naive Bayes, multinomial Naive Bayes, Bernoulli Naïve Bayes, random forest, and perceptron classifier). An ensemble majority-voting classifier has been culminated, so as to compile the results from all the individual classifiers and algorithms, and on comparing the results we can deduce that the proposed machine learning classifier is really apt for a network intrusion detection system for 802.11 networks, data for which has been captured in the form of packets in the AWID-(II)-2019 dataset. The

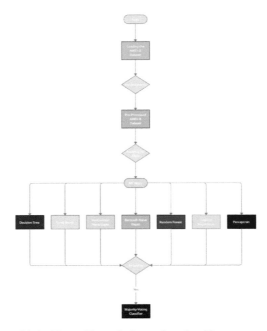

Figure 11.4   Ensemble majority-voting classifier.

voting classifier used here is the majority-hard voting-based classifier wherein the input data is classified based on the mode of the predictions by all of the machine learning classifiers at an individual level. Majority hard voting classifier works on the principal of majority-voting wherein it's needed to consider an odd number of classifiers to train and test the data with. This paper considers seven classifiers (and odd number) to make the ensemble, and the working is basically that if 4 out of 7 classifiers predict the network traffic flow as normal, the final class assigned to the traffic will be the "normal" class and if 4 out of 7 classifiers predict traffic as abnormal, the final class assigned to the traffic will be the "abnormal" class. The proposed classifier has been programmed as a Python-scripted code including the application of all the above stated classifiers, to gauge their intrusion detection-based classification capabilities, in terms of the accuracy they achieve while being applied over the AWID-(II)-2019 dataset.

11.4 Results and Discussion

The proposed ensemble majority-voting classifier was scripted in Python on a system with Mac OS Big Sur, having a computational memory of 16GB, 2.4 GHz Clock Frequency and Intel UHD Graphics GPU. This summarizes the system environment on which the research and experiments were conducted, so as to deduce the ensemble majority-voting classifier. On application of all the classifiers, that is using the training set to train the models and then using the testing set on the model to check its prediction capability. The accuracies of each classifier have been compared, the results of which are depicted in the Table 11.2. It is observed that the highest accuracy is obtained by the majority-voting classifier, and the least accuracy is obtained by the multinomial Naïve Bayes algorithm. The highest and lowest accuracies are 90.45% and 70.86% respectively, thereby proving that the majority-voting

Table 11.2 Comparative accuracies of various classifiers with majority-voting classifier.

Serial No	Classifier	Accuracy
1	Decision tree	90.44%
2	Gaussian Naive Bayes	74.77%
3	Multinomial Naive Bayes	70.86%
4	Random forest	90.44%
5	Logistic regression	90.43%
6	Perceptron	85.51%
7	Majority-voting classifier	90.45%

classifier has given the best results, in terms of its capability to classify a given network traffic flow as normal or abnormal.

11.5 Conclusion

The main objective of this paper was to conduct an extensive and comprehensive literature review of pre-existent machine learning-based classifiers for 6G wireless network security and proposes new ensemble majority-voting classifier for detecting cyberattacks in 6G wireless networks. The dataset being considered for research purposes in this paper was the AWID-(II)-2019 dataset, which can include cyber intrustion details such as *arp*, *amok*, *beacon*, *deauthentication*, *authentication_request*, *cafe_latte*, *fragmentation*, *probe_response*, and *evil_twin*, by classifying and bifurcating the network flow into the following two categories, that is "normal" and "abnormal." After preprocessing the dataset and reducing the number of columns from 155 to 20, the network traffic bifurcation takes place. Following this, the various classifiers and models have been trained using the reduced training set and then exposed to the reduced test set of data. On testing and comparing, it is concluded that the proposed ensemble majority-voting classifier gives the highest accuracy of 90.45%, which can be considered to be a really good accuracy accounting for the number of tuples in the considered dataset, and the least accuracy of 70.86% is given by the multinomial Naive Bayes classifier. Thus, the proposed ensemble majority-voting classifier can be considered as a good alternative to pre-existent works by other researchers and scholars, in terms of its ability to predict whether an incoming traffic wireless message belongs to a normal or an abnormal/malicious network flow. In future, the work will concentrate on exploring deep learning techniques for cyberattack detections and further the work will also concentrate on building own dataset and developing model for it.

References

[1] Yu-Feng H, Chia-Ying L, Wei-Yang L. "Intrusion detection by machine learning: A review". Expert Systems with Applications 36 (2009) 11994–12000. DOI:10.1016/j.eswa.2009.05.029

[2] Suad O, Fadl M. B, Nabeel A and Amal A. "Intrusion detection model using machine learning algorithm on Big Data environment". Othman et al. J Big Data (2018) 5:34 https://doi.org/10.1186/s40537-018-0145-4

[3] C. Kolias, G. Kambourakis, A. Stavrou and S. Gritzalis, "Intrusion Detection in 802.11 Networks: Empirical Evaluation of Threats and a Public Dataset," in IEEE Communications Surveys & Tutorials, vol. 18, no. 1, pp. 184-208, Firstquarter 2016, doi:10.1109/COMST. 2015.2402161

[4] AWID-(II)-2019 Dataset, https://icsdweb.aegean.gr/awid/awid2. Accessed 23 February 2022.

[5] Suad O, Fadl M. B, Nabeel A and Amal A. "Intrusion detection model using machine learning algorithm on Big Data environment". Othman et al. J Big Data (2018) 5:34 https://doi.org/10.1186/s40537-018-0145-4

[6] Kamran S, Suhuai L, Shan C and Dongxi L. "Cyber Threat Detection Using Machine Learning Techniques: A Performance Evaluation Perspective". IEEE International Conference on Cyber Warfare and Security DOI: 10.1109/ICCWS48432.2020.9292388

[7] Rishabh D and Thomas M. "Machine Learning and Cyber Security". 2017 International Conference on Computer, Electrical & Communication Engineering (ICCECE).DOI:10.1109/ICCECE.2017.8526232

[8] Razan A, Miad F, Abdelshakour A and Arafat A. "Deep and Machine Learning Approaches for Anomaly-Based Intrusion Detection of Imbalanced Network Traffic". January 2019. IEEE Sensors Letters 3(1):1-4. DOI: 10.1109/LSENS.2018.2879990

[9] Yuyang Z, Guang C, Shanqing J, Mian D. "Building an Efficient Intrusion Detection System Based on Feature Selection and Ensemble Classifier". June 2020 Computer Networks 174. DOI: https://doi.org/10.1016/j.comnet.2020.107247.

[10] Abrar, Iram & Ayub, Zahrah & Masoodi, Faheem & Bamhdi, Alwi. (2020). A Machine Learning Approach for Intrusion Detection System on NSL-KDD Dataset. 919-924. 10.1109/ICOSEC49089.2020 .9215232.

[11] Ahmad, T. & Aziz, Moh Nasrul. (2019). Data preprocessing and feature selection for machine learning intrusion detection systems. ICIC Express Letters. 13. 93-101. 10.24507/icicel.13.02.93.

Biographies

Karan Sharma received his Bachelor of Engineering degree in computer science engineering from the Birla Institute of Technology and Science, Pilani, Dubai in 2022. He has been awarded with the Director's Gold Medal for All-Round Achievement for his Excellence in Academics and Co-Curriculars. He was the President of the IEEE Society for Students and the President of the Microsoft Tech Club while his tenure at BITS, and worked on numerous research and design projects revolving around the field of cybersecurity, cyber threat intelligence, machine learning, and artificial intelligence.

His research interests are based around finding innovative solutions for cyber threat detection and detection of cyberattacks using applied machine learning classifiers, artificial intelligence models, and leveraging neural networks and blockchain-based techniques for cybersecurity management.

Karan is currently working at PwC Middle East, as a consultant in the Microsoft Practice, working on digital transformation and automation projects for various clients leveraging the Microsoft Tech Stack and developing innovative solutions. He has acquired numerous certifications by Microsoft namely PL-900, PL-100, PL-200, MB-210, MB-220, and MB-230 alongside cybersecurity-based certifications namely the ISO Lead Auditor Certifications for ISO 27001:2013, ISO 22301:2019, and ISO 20000-1:2018.

Kavya Parthasarathy received her Bachelor of Engineering degree in computer science from the prestigious Birla Institute of Technology and Science Pilani, Dubai in 2022. Her journey through academia and appetite for learning, coupled with unwavering commitment, earned her the coveted Chancellor's Silver Medal for academic excellence. She has a profound connection with the enigmatic realms of cybersecurity and data privacy.

From the corridors of her academic institution to the echelons of certification excellence, she aims to master the art of securing digital landscapes.

To further her journey in the domain of data privacy, she has attained the Certified Information Privacy Professional/Europe (CIPP/E) certification that encompasses pan-European and national data protection laws, practical concepts concerning the protection of personal data, and trans-border data flows, thereby demonstrating comprehensive General Data Protection Regulation (GDPR) knowledge, perspective, and understanding to ensure compliance and data protection success. She has also achieved the Lead Implementer certification for ISO 27701:2019 marking a commitment to understanding and ensuring the privacy of data subjects' personally identifiable information (PII).

She is expanding her knowledge and expertise within the realm of cybersecurity by successfully clearing the Lead Auditor and Lead Implementer certifications for ISO 27001:2022 (Information Security Management System), ISO 22301:2019 (Business Continuity Management System), and ISO 20000-1:2018 (IT Service Management System) which showcases her ability to not only comprehend the intricacies of these standards but also translate it into reality through control implementation.

Beyond the realm of certificates and accolades, she is currently pursuing her career in the field of governance, risk, and compliance which aims at ensuring that an organization operates in a secure and compliant manner while effectively managing and mitigating risks. Her research interests lie at the dynamic intersection of cybersecurity, particularly, around the domain of threat detection in an effort to devise innovative strategies, solutions and robust threat detection mechanisms that pre-emptively identify and neutralize cyber threats in order to fortify digital ecosystems against emerging risks.

Pranav M. Pawar graduated in computer engineering from Dr. Babasaheb Ambedkar Technological University, Maharashtra, India, in 2005, received a Master in computer engineering from Pune University, in 2007, and received Ph.D. degree in wireless communication from Aalborg University, Denmark in 2016; his Ph.D. thesis received a nomination for Best Thesis Award from Aalborg University, Denmark. Currently, he is working as an assistant professor in the Department of Computer Science, Birla Institute of Technology and Science, Dubai. Before BITS, he was a postdoctoral fellow at Bar-Ilan University, Israel from March 2019 to October 2020 in the area of wireless communication and deep leaning. He is the recipient of an outstanding postdoctoral fellowship from the Israel Planning and Budgeting Committee. He worked as an associate professor at MIT ADT University, Pune from 2018 to 2019 and also as an associate professor in the Department of Information Technology, STES's Smt. Kashibai Navale College of Engineering, Pune from 2008 to 2018. References 15 from 2006 to 2007, was working as system executive in POS-IPC, Pune, India. He received recognition from Infosys Technologies Ltd. for his contribution to the Campus Connect Program and also received different funding for research and attending conferences at the international level. He published more than 40 papers at the national and international levels. He is **IBM DB2** and **IBM RAD** certified professional and completed NPTEL certification in different subjects. His research interests are energy-efficient MAC for WSN, QoS in WSN, wireless security, green technology, computer architecture, database management system, and bioinformatics.

Raja Muthalagu is currently an associate professor with Birla Institute of Technology and Science, Pilani, Dubai Campus, Dubai, USA. He was a postdoctoral research fellow with Air Traffic Management Research Institute, Nanyang Technological University, Singapore, from 2014 to 2015. Dr. Muthalagu was the recipient of the Canadian Commonwealth Scholarship Award 2010 for the Graduate Student Exchange Program in the Department of Electrical and Computer Engineering, University of Saskatchewan, Saskatoon, SK, Canada. His research interest includes wireless communication.

12

From Connectivity to Intelligence: Integrating IoT-6G for the Future

Tuhina Raj, Iqra Javid, Madhukar Deshmukh, and Sibaram Khara

Sharda University, India

Abstract

The integration of Internet of Things (IoT) and sixth generation (6G) wireless communication networks are expected to redefine customer services and applications and pave the way for completely intelligent and autonomous systems in the future. It has the potential to revolutionize connectivity, communication, and data sharing. The network demands have been increased due to evolving technology. Ultra-high data rates and extremely low latency are required. With a 1 Tbps data rate, less than 1 ms latency, high throughput, high level of security, and the capacity to link trillions of objects, 6G integrated with IoT has the potential to offer many beneficial opportunities and advantages. IoT devices will gain significantly more connectivity by utilizing the faster speeds and lower latency of 6G, enabling almost instantaneous data transfer and real-time communication. Additionally, the enormous number of networked devices, sensors, and actuators supported by 6G's large deployment capability will strengthen IoT ecosystems.

Keywords: Machine-to-Machine Communication, Latency, Internet of Things, Mobility, 6G

12.1 Introduction

The integration of Internet of Things (IoT) and 6G holds immense potential and is poised to drive significant advancements and developments across various sectors. IoT and 6G integration is a revolutionary combination that can open various untapped possibilities. IoT networks could expand exponentially and include a wide range of devices and systems. This explosion of connection makes it possible for devices to connect with one another and exchange data seamlessly, which optimizes decision-making processes and upgrades efficiency. Owing to the IoT's multiple uses in both commercial and human-centric applications the globe is focusing more on it nowadays. However, the actual applications showcasing the general benefits of the IoT are based on device-to-device (D2D), machine-to-machine (M2M), and vehicle-to-vehicle (V2V)/V2X communication technologies [1, 6]. The Industrial Internet of Everything (IIoE) evolved because of the industrial control systems (ICSs) being integrated with the Internet of Everything (IoE), which offers outstanding solutions for enormous data transmission to the edge network [7]. An additional significant challenge to the development of effective IoT applications is the reliable and low-latency data transfer [8]. Owing to the exceptional features over the prior network generations, 6G is anticipated to provide an entirely new quality of service and enhance user's experiences in IoT systems [9, 10]. These previously unheard-of levels of capacity will hasten the applications and deployments of 6G-based IoT

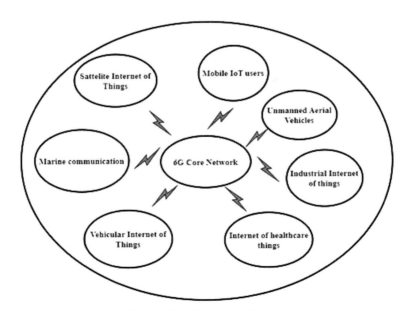

Figure 12.1   Integrated 6G-IoT.

Table 12.1 Parameter analysis for 4G, 5G, and 6G wireless communication technology.

Features	4G	5G	6G
Speed	Up to 100 Mbps	Up to 10 Gbps	Expected to exceed 100 Gbps
Latency	Around 50 ms	Around 1 ms	Expected to be ultra-low
Capacity	Limited bandwidth	Enhanced capacity	Massive capacity
Connection density	Upto 2k devices/km^2	Upto 10^5 devices/km^2	Expected to support even higher density
Spectrum efficiency	Moderate efficiency	High efficiency	Expected to be even more efficient
Network slicing	Not supported	Supported	Advanced support expected
Internet of Things (IoT)	Limited IoT support	Extensive IoT support	Enhanced IoT capabilities
Energy efficiency	Improved compared to 3G	Further improvements	Expected to be highly energy-efficient
Applications	Mobile broadband, voice, video streaming	AR/VR, autonomous vehicles, industrial automation	AI-powered applications, holographic communication, advanced IoT use cases

networks in the areas of wireless communication, IoT data sensing, device connection, and 6G network management. Table 12.1 shows the detailed analysis for the recent generations of wireless communication generations and enhances the aspect as in why the researchers have been keen to integrate the IoT along with it.

6G, an evolutionary generation, will revolutionize human life and society by offering superior network performance, higher data rates, lower latency, and three-dimensional coverage. It will also provide more accurate localization, stricter privacy, and security. 6G's multitudinous use cases will result in different and conflicting requirements, with several trends emerging. 6G enables massive interconnectivity in IoT with diverse service requirements, providing a new network architecture and breakthrough technologies to connect millions of devices and applications seamlessly ensuring performance guaranteed [11]. Figure 12.1 shows the integration of 6G with IoT.

Table 12.2 Comparison of 5G-IoT vs. 6G-IoT [13] – Features.

5G-IoT	6G-IoT
Data speed of 20 Gbps	Data speed from 1 Tbps
Mobility of the traffic: 10 Mbps/m^2	Mobility of the traffic: 1 Gbps/m^2
Connectivity density: 10^6 devices/km^2	Connectivity density: 10^7 devices/km^2
Network latency: 1 ms	Network latency: 10–100 μs
Coverage percentage: about 70%	Coverage percentage: >99%
Energy efficiency: 1000× relative to 4G	Energy efficiency: 10× relative to 5G
Spectrum efficiency: 3–5× relative to 4G	Spectrum efficiency: >3× relative to 5G

12.2 Features of Integrated 6G-IoT

As compared to its 5G predecessor, 6G is anticipated to provide new innovative wireless technologies and inventive networking infrastructures to achieve several new IoT applications while fully satisfying such stringent network requirements [12]. With the development of cutting-edge technologies like edge intelligence, THz, and large-scale satellite constellation, 6G communication systems can contribute towards a more robust IoT ecosystem. It can create an entirely associated and intelligent digital world towards the planned economic, social, and environmental ecosystems. The comparison of 5G-IoT and 6G-IoT is shown in Table 12.2. The 6G-IoT is expected to perform better than 5G-IoT in a variety of aspects. While 6G networks can offer a higher standard of network characteristics with the notable characteristics, 5G networks are still able to experience certain severe limitations regarding the availability of cellular congestion performance, capacity of connectivity between devices, and latency within the network.

12.3 Enabling 6G Technologies for IoT

The various enabling technologies [14] are shown in Figure 12.2.

12.3.1 Massive URLLC (ultra-reliable low-latency communication)

This key technology plays a vital role in the integration of 6G and IoT devices. These techniques support ultra-reliable and low-latency requirements for the communication. Real-time data transmission and low latency is becoming increasingly important for real-time and high-end applications, especially in healthcare where it plays a crucial role in emergency situations.

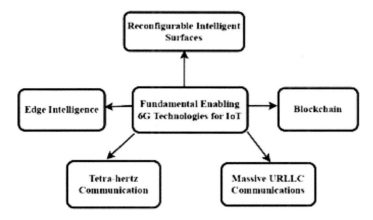

Figure 12.2   Enabling technologies.

12.3.2 Terahertz communications

This novel technology uses frequencies in the range of 0.1–10 THz for wireless communications. This technology helps in achieving high data rate when compared to the established conventional wireless technologies which makes a promising alternative for bandwidth demand of Internet of Things.

Billions of Internets of Things devices can gather and analyze data in real time via terahertz communications.

12.3.3 Blockchain

The Internet of Things (IoT) can be significantly facilitated by 6G technologies, which include blockchain technology. For IoT devices, it may provide a decentralized identity management system and includes security features like immutability, decentralized consensus, and cryptographic algorithms. Automation of procedures like device management, data exchange and service delivery can be made possible by blockchain technology in 6G networks.

12.3.4 Edge Intelligence

To provide real-time decision-making and swift responses, edge intelligence is crucial for 6G technologies for the Internet of Things. It improves data privacy and security, decreases latency, and maximizes bandwidth utilization.

Table 12.3   Enabling technologies for 5G-IoT and 6G IoT.

Enabling technologies	5G-IoT	6G-IoT
Terahertz communications	Not supported	Supported and optimized
Edge intelligence	Limited integration with IoT	Advanced integration and capabilities
Reconfigurable intelligent surfaces	Not supported	Supported for enhanced connectivity
Blockchain	Limited integration for security	Enhanced integration for trust and security
Massive URLL communications	Supported, but with limitations	Enhanced support and reliability

It enhances overall network efficiency and assists in reducing congestion in the network. In addition to enabling offline operation and resilience, real-time analytics, and insight creation, and lowering energy consumption in IoT installations, edge intelligence lowers the risk of data breaches and unauthorized access to sensitive data. It is especially helpful for IoT devices that run on batteries since it increases their battery life and lessens how frequently they need to be recharged or replaced.

12.3.5 Reconfigurable intelligence surface

For IoT, reconfigurable intelligent surfaces (RIS) have the potential to be a key enabler of 6G technologies. Through optimizing signal propagation and lowering transmit power, RIS may improve wireless coverage, expand network capacity, and support energy-efficient IoT communications. RIS may also improve security and privacy, enabling intelligent resource allocation, and help IoT devices accurately localize and position themselves on 6G networks. Table 12.3 compares the enabling technologies for 5G-IoT and 6G-IoT.

12.4 IoT Applications for 6G

6G is being used in a variety of significant IoT fields, such as Healthcare Internet of Things, Vehicular Internet of Things, self-driving UAVs, Satellite Internet of Things, and the Industrial Internet of Things. Thanks to the incorporation of 6G as well as associated technologies, the Healthcare Internet of Things (HIoT) [15] will see a revolution. According to the research in [16], 6G technologies like mURLLC and THz communications might be employed to speed up medical network connections between wearable devices and far-off physicians and enable exceptionally low-latency healthcare data transfer. In fact, healthcare domains like remote health monitoring require a low-rate communication (below 1 ms) with a dependability requirement of over 99.999% of the order to give real-time health provision with an immediate and accurate diagnosis at a distance. It's relevant to note that 6G technology may be utilized to do surgery far away, enabling remote surgeons to control the process by utilizing robotic equipment with a millisecond's latency and high reliability. A recent study in [17] also looked at a remote surgical system that uses UAVs and blockchain in the context of 6G. In an emergency [18], light medical items like drugs and surgical equipment are sent between hospitals using unmanned aerial vehicles as relays to address the long wait periods for medical assistance. This reduces data transmission latency and minimizes traffic jams on the roads. In 6G-based health-related networks, the transmission of information will demand ultra-high data speeds in the future, and mURLLC technology is the ideal solution [19]. Consequently, small devices, implants, and on-body sensing devices may transport real-time data to edge devices with an exceptionally high level of availability and confidence for short- and long-term medical analysis [20]. Additionally, by enabling reasonably fast (up to 100 km/h) real-time video streaming with good color resolution for accurate diagnosis to physicians and paramedical personnel from the hospital, 6G-based URLLC can be used to support linked ambulance in future healthcare [21]. Given the security risks in current mobile surgery networks, blockchain has been integrated into the robotic framework where every robotic device act as a data node to allow for the safe storage of surgical information in the database ledger without the need for established authority. More specifically, electronic agreements are being implemented that can give control over the exchange of health data during surgery and automated verification for requests for health data.

Artificial intelligence may be utilized for data learning and analytics to realize intelligent 6G-based healthcare. The study in [22] analyzes historical health data of stroke out-patients obtained from wearable sensors in healthcare-based 6G heterogeneous networks using machine learning (ML) techniques. An uplink radio resource allocation optimization solution is incorporated, where the resources assigned are in proportion to the patient's risk of stroke and to speed up the stroke care for patients.

Another suggestion for using machine learning (ML) in intelligent 6G health networks is made by [23] and suggested edge cloud infrastructure to deliver low-latency health data analytics for medical services. The rapid global spread of COVID-19 in recent years has caused extremely serious health concerns for several countries.

As a result of the 6G network's implementation of detection and communication, mobile robots will be capable of operating together quickly and reliably [24] increasing manufacturing's efficiency, precision, and adaptability [25]. The well-being of operators is intimately correlated with the stability and security of automotive communication, even though autonomous cars have distinct platforms. A basic infrastructure for vehicle communication devices might be provided by 6G networks. Autonomous automobiles may communicate information with servers or other vehicles thanks to 6G networks' improved ultra-reliable low-latency connection, which improves road safety [26].

In lieu of using the nervous system of humans, brain-computer interface (BCI) technologies offer substitute means of communication and control [27–29]. The development of wireless BCI is a result of the evolution of mobile communication technology.

By utilizing high-performance mobile networks, extended reality (XR) 6G can encourage the growth of XR technologies in the IoT space. Virtual reality and augmented reality both fall under the umbrella of extended reality, which mixes the actual world with virtual settings. While AR uses mobile devices to add digital material to the actual world, VR creates a computer-simulated virtual environment utilizing a headgear that produces sounds and visuals [27]. The establishment of 6G networks, which provide dependable Internet access, fast connectivity data transfer rates, as well as low is anticipated to be advantageous for both VR and AR. VR technology will be able to maximize mobility without wires thanks to 6G's ultra-mobile broadband service.

12.5 Benefits of Integrated 6G-IoT

The IoT involves connecting various physical devices and objects to the Internet, enabling them to communicate and share data with each other. While the fifth generation of wireless technology (5G) has already brought significant advancements to IoT applications, the upcoming 6G is expected to further enhance the capabilities and benefits of IoT. Here are some potential benefits of IoT in 6G:

1. Effective interaction: the Internet of Things (IoT) and 6G enable machine-to-machine (M2M) communication, a capacity to link physical

objects, full availability of information with minimizing inconsistencies, and superior performance [28].

2. Effective technology and management: wireless technology connects all physical equipment, allowing them to share digital data resources. Powerful control and automation systems handle the entire situation. The machine can communicate with others without human involvement, resulting in quicker and more immediate results [29].

3. Enhanced business decision-making is made possible thorough monitoring and information discovery system with cutting-edge knowledge capabilities. It enables more accurate monitoring and control of product expiration, access to data which was previously impossible, and exact tracking of supply quality and quantity.

4. IoT is highly beneficial for individuals in their daily lives because it makes electrical gadgets smarter so they can interact effectively with one another and save time, money, and energy. Speedier and the adoption of new standards can automate daily chores effortlessly using IoT, such as in smart offices, and execute complicated jobs more quickly and effectively [30]. According to the latest requirements, the sophisticated gadget automatically upgrades the system.

5. 6G is likely to leverage edge computing capabilities, bringing computational power closer to the IoT devices themselves. This would enable real-time data processing, reduced latency, and distributed intelligence, allowing IoT devices to make quick and autonomous decisions locally without relying solely on centralized cloud resources.

6G is expected to offer extended communication ranges, enabling IoT devices to connect and communicate over longer distances. This would be particularly beneficial for IoT deployments in rural areas, remote locations, or large-scale industrial environments where devices are widely distributed.

12.6 Limitations of Integrated 6G-IoT

While the Internet of Things (IoT) has several advantages, it's important to consider the potential disadvantages it may bring when combined with 6G technology. Sensitive data must be protected and unauthorized access must be avoided by using adequate security measures and encryption solutions. Data is generated and collected by IoT devices, but greater connection and data interchange brought about by 6G present privacy issues. To preserve user privacy, strict privacy laws and data protection measures are required.

To integrate IoT in 6G, a strong and wide infrastructure is needed, which might be expensive and difficult in rural or undeveloped places. To ensure interoperability and simple integration among devices and networks, the Internet of Things (IoT) community must make efforts to develop shared standards and protocols. To minimize environmental impact and properly control energy consumption, IoT devices and infrastructure must be energy efficient. This will lower the need for energy in 6G. For IoT and 6G data to yield beneficial conclusions and avoid data overload, appropriate data management along with processing technologies must be in place.

12.7 Research Scope and Challenges

The effective management of large-scale installations of Internet of Things devices in a 6G network could be a challenge in many aspects.

In comparison to previous generations, 6G is anticipated to accommodate a substantially higher number of IoT devices because it intends to offer unprecedented speeds, extremely low latency, and huge connection. However, maintaining and administering a sizable number of IoT devices can present several problems including allocation of resources, flexibility, and congestion in the network.

These specific areas for research inquiry include:

- Investigating methods to scale the network infrastructure so that it can support the enormous number of IoT devices in a 6G environment. To address the growing device density, this may need creating novel protocols, network structures, or resource management strategies.
- Exploring power management techniques and energy-efficient communication mechanisms for IoT devices on a 6G network. When dealing with a large-scale deployment, efficient power utilization is essential to ensuring extended battery life and sustainable operations for IoT devices.
- Studying methods to deliver reliable and predictable Quality of Service (QoS) for various IoT applications and services over a 6G network. For the success of crucial IoT applications like autonomous smart grids, and healthcare systems, it is crucial to guarantee low latency, high throughput, and uninterrupted connectivity.
- Addressing the particular security and privacy issues brought on by the combination of IoT and 6G networks. To safeguard IoT devices, data, and communications from possible hacking attempts, strong authentication, encryption, and access control systems are being developed.

- Investigating effective methods for data storage, processing, and analytics to manage the enormous amount of data produced by IoT devices on a 6G network. To reduce latency and maximize network resources, researchers are looking into the use of edge computing, shared storage, and data aggregating techniques.
- Studying standards and protocols that enable seamless connectivity across various IoT devices and 6G networks constitutes standardization and interoperability developing frameworks that guarantee interoperability, simplicity of integration, and easy communication between heterogeneous devices and networks.

12.8 Conclusion

The integration of the Internet of Things (IoT) and 6G technologies holds great potential for driving advancements in connectivity, data processing, and communication systems. It can enable seamless connectivity and communication between a vast array of devices, sensors, and systems, facilitating real-time data exchange and decision-making. However, the integration of IoT and 6G presents challenges such as addressing security and privacy concerns, managing the complexity of interconnected systems, ensuring interoperability across different devices and networks, and handling the massive volumes of data generated by IoT devices. To fully realize the potential of integrating IoT and 6G technologies, collaboration among industry stakeholders, researchers, policymakers, and standardization bodies is crucial. Efforts should focus on developing robust security measures, establishing interoperability standards, optimizing network infrastructure, and addressing ethical considerations in the use of IoT and 6G technologies.

References

[1] P. Pawar, A. Trivedi, "Device-to-device communication based IoT system: benefits and challenges," IETE Technical Review 36, pp. 362-374, 2019.
[2] AM. Rahmani, et al., "Internet of things applications: opportunities and threats," Wireless Personal Communications, pp. 451-476, 2022.
[3] PK. Padhiand, F. Charrua-Santos, "6G enabled industrial internet of everything: Towards a theoretical framework," Applied System Innovation, 2021.

[4] OA. Amodu, M. Othman, "Machine-to-machine communication: An overview of opportunities," Computer Networks, pp. 255-276, 2018.

[5] M El. Zorkany, et al.,"Vehicle to vehicle "V2V" communication: scope, importance, challenges, research directions and future," The Open Transportation Journal, 2020.

[6] S. Iqbal, et al., "Efficient IoT-based formal model for vehicle-life interaction in VANETs using VDM-SL," Energies, 2022.

[7] BB. Gupta, M. Quamara, "An overview of Internet of Things (IoT): Architectural aspects, challenges, and protocols," Concurrency and Computation: Practice and Experience, 2020.

[8] MZ. Khan, et al., "Reliable Internet of Things: Challenges and future trends," Electronics, 2021.

[9] L. Bariah et al., "A prospective look: Key enabling technologies, applications and open research topics in 6G networks," IEEE access, pp. 174792-174820, 2020.

[10] X. You, et al., "Towards 6G wireless communication networks: Vision, enabling technologies, and new paradigm shifts," Science China Information Sciences, 2021.

[11] F. Guo, et al., "Enabling massive IoT toward 6G: A comprehensive survey," IEEE Internet of Things Journal, pp. 11891-11915, 2021.

[12] A. Shahraki, "A comprehensive survey on 6G networks: Applications, core services, enabling technologies, and future challenges," arXiv preprint arXiv:2101.12475, 2021.

[13] DC. Nguyen, "6G Internet of Things: A comprehensive survey," IEEE Internet of Things Journal, pp. 359-383, 2021.

[14] DC.Nguyen, et al.,"6G Internet of Things: A comprehensive survey," IEEE Internet of Things Journal, pp. 359-383, 2021

[15] H. Habibzadeh, et al., "A Survey of Healthcare Internet of Things (HIoT): A Clinical Perspective," IEEE Internet of Things Journal, pp. 53–71, Jan. 2020.

[16] S. Nayak, R. Patgiri, "6G Communication Technology: A Vision on Intelligent Healthcare," pp. 1–18, 2021.

[17] DC. Nguyen, et al., "Blockchain and Edge Computing for Decentralized EMRs Sharing in Federated Healthcare," in Proceedings of the 2020 IEEE Global Communications Conference, Taipei, Taiwan, pp. 1–6, Dec. 2020.

[18] L. Mucchi, et al., "How 6G Technology Can Change the Future Wireless Healthcare," in Proceedings of the 2nd 6G Wireless Summit (6G SUMMIT), Levi, Finland, pp. 1–6, 2020

[19] M. S. Kaiser, et al., "6G Access Network for Intelligent Internet of Healthcare Things: Opportunity, Challenges, and Research Directions," in Proceedings of International Conference on Trends in Computational and Cognitive Engineering, pp. 317–328, Singapore, 2021.

[20] G. Cisotto, et al., "Requirements and Enablers of Advanced Healthcare Services over Future Cellular Systems," IEEE Communications Magazine, vol. 58, no. 3, pp. 76–81, Mar. 2020.

[21] M. S. Hadi, et al., "Patient-Centric HetNets Powered by Machine Learning and Big Data Analytics for 6G Networks," IEEE Access, vol. 8, pp. 85 639–85 655, 2020.

[22] A. H. Sodhro, et al., "Towards ML-based Energy-Efficient Mechanism for 6G Enabled Industrial Network in Box Systems," IEEE Transactions on Industrial Informatics, pp. 1–1, 2020.

[23] Y. Siriwardhana, et al., "The role of 5G for digital healthcare against COVID-19 pandemic: Opportunities and challenges," ICT Express, Nov. 2020.

[24] W. Saad, et al., "A vision of 6G wireless systems: Applications, trends, technologies, and open research problems," IEEE network, 34(3), pp.134-142. 2019.

[25] J.F. Monserrat, "Key technologies for the advent of the 6G," In 2020 IEEE wireless communications and networking Conference Workshops (WCNCW), pp. 1-6, April 2020.

[26] J.H. Kim, "6G and Internet of Things: a survey," Journal of Management Analytics, pp.316-332, 2021.

[27] O. Danielsson, et al., "Augmented reality smart glasses in industrial assembly: Current status and future challenges," Journal of Industrial Information Integration, 2020.

[28] T. Aditya, "Internet of things (IoT): research, architectures and applications," Int J Futur Revol Comput Sci Commun Eng 4(3): pp. 23–27, 2018.

[29] F. Ciro, et al., "The advantages of IoT and cloud applied to smart cities," 3rd international conference on future internet of things and cloud. IEEE, pp. 325–332, 2015.

[30] BD.Deebak, et al.,"Drone of IoT in 6G wirelesscommunications: technology, challenges, and future aspects," Unmanned aerial vehicles in smart cities. Cham: Springer, pp. 153–165, 2020.

Biographies

Tuhina Raj is currently pursuing her M.Tech. in digital communication from Sharda University, India. She has obtained her B.E. in electronics and communication engineering from Anna University, India. Her research interest includes wireless networks, human activity recognition, and antennas.

Iqra Javid is currently pursuing her Ph.D. in wireless communication from Sharda University, India. She has obtained her M.Tech. in digital communication from Sharda University, India and her B.E. in electronics and communication engineering from SSM College of Engineering and Technology Pattan, Jammu and Kashmir, India. Her research interest includes wireless networks and device-to-device communications.

Madhukar Deshmukh received his Master's degree from MNNIT, Allahabad, India, and his Ph.D. degree in wireless communication from Aalborg University, Denmark, specifically in the area of signal processing for cognitive communication and the networks. He received a postdoctoral fellowship from the Israeli Science Foundation and worked with Prof. Amir Leshem at Bar-Ilan University on wireless networks. His research interests are signal processing and wireless and cognitive communication. He is a reviewer on IEEE systems journals, Springer's WPC and is on review committees of IEEE conferences such as IEEE International Black Sea Conference on Communications and Networking, IEEE WPMC, etc. He has a lengthy teaching and research experience.

 **Sibaram khara** received his Ph.D. in engineering from Jadavpur University, Kolkata, in next-generation wireless heterogeneous networks — essentially in the area of interworking network and protocol convergence techniques for cellular and Wi-Fi integrated networks. He did PG in Digital Systems from National Institute of Technology, Allahabad. He received Best Paper awards for his analytical model of cellular/Wi-Fi system (IEEE ADCOM 2008 MIT Chennai and IEEE EWT 2004 (1st), I2IT Pune (3rd)). He was honored as best research faculty in the School of Electronics Engineering, VIT University, Vellore, India for year 2010. His research articles are presented at seminars and conferences in many countries, namely, WEAS02 Athens, IEEE VTC06 Melbourne, IEEE PWC07 Prague, IEEE/ACM SAC10 Switzerland, etc. His major research interests cover the areas of cluster-based wireless sensor networks, spectrum mobility in cognitive radio system, call admission control in heterogeneous network, and carrier aggregation in LTE-A technology.

Index

About the Editors

Prof. Dr. Ramjee Prasad, Fellow IEEE, IET, IETE, and WWRF, is a Professor Emeritus of Future Technologies for Business Ecosystem Innovation (FT4BI) in the Department of Business Development and Technology, Aarhus University, Herning, Denmark. He is the Founder President of the CTIF Global Capsule (CGC). He is also the Founder Chairman of the Global ICT Standardization Forum for India, established in 2009. He has been honored by the University of Rome "Tor Vergata," Italy as a Distinguished Professor of the Department of Clinical Sciences and Translational Medicine on March 15, 2016. He is an Honorary Professor at the University of Cape Town, South Africa, and the University of KwaZulu-Natal, South Africa, and also an Adjunct Professor at Birsa Institute of Technology, Sindri, Jharkhand, India. He has received Pravasi Bhartiya Samman Puraskaar (Emigrant Indian Honor Award by the Indian President) on January 10, 2023 in Indore. He is recipient of the prestigious Distinguished Alumni Award under the category 'Excellence in Teaching and Research in Engineering and Technology' for 2023, by the Birla Institute of Technology, Mesra, Ranchi MESRA, RANCHI, Jharkhand, India. He has received "Pioneering Visionary Award" by Bihar Jharkhand Association of North America (BAJANA) at New Jersey, USA, on Sunday March 26, 2023.He has received the Ridderkorset of Dannebrogordenen (Knight of the Dannebrog) in 2010 from the Danish Queen for the internationalization of top-class telecommunication research and education. He has received several international awards such as the IEEE Communications Society Wireless Communications Technical Committee Recognition Award in 2003 for making a contribution in the field of "Personal, Wireless and Mobile Systems and Networks," Telenor's Research Award in 2005 for impressive merits, both academic and organiza- tional within the field of wireless and personal communication, 2014 IEEE AESS Outstanding Organizational Leadership Award for: "Organizational Leadership in developing and globalizing the CTIF (Center for TeleInFras- truktur) Research Network," and so on. He has been the Project Coordinator of several EC projects, namely, MAGNET, MAGNET Beyond, eWALL. He has published more than 50 books, 1000 plus journal and conference publications, more than 15 patents, over 155 Ph.D. Graduates and a larger number of Masters (over 250). Several of his students are today's worldwide telecommunication leaders themselves.

 **Dnyaneshwar Shriranglal Mantri** graduated in Electronics Engineering from Walchand Institute of Technology, Solapur (MS), India in 1992 and received Master's from Shivaji University in 2006. He has been awarded Ph.D. in Wireless Communication at the Center for Tele InFrastruktur (CTIF), Aalborg University; Denmark. He has teaching experience of 25 years.

From 1993 to 2006, he was working as a lecturer in different institutes [MCE Nilanga, MGM Nanded, and STB College of Engg. Tuljapur (MS) India].

Since 2006, he has been associated with Sinhgad Institute of Technology, Lonavala, Pune and presently he is working as a professor with the Department of Electronics and Telecommunication Engineering. He is a recognized master's and Ph.D. guide at Savitribai Phule Pune University, Pune in the subjects of electronics and telecommunication engineering, computer and information technology. He is a Senior Member of IEEE, a Fellow of IETE, and Life Member of ISTE. He has published 06 books, 20 journal papers in indexed and reputed journals (Springer, Elsevier, IEEE, etc.), and 19 papers in IEEE conferences. He is a reviewer of international journals (Wireless Personal Communication, Springer, Elsevier, IEEE, Communication Society, MDPI, etc.) and conferences organized by IEEE. He worked as a TPC member for various IEEE conferences and also organized IEEE conferences GCWCN2014 and GCWCN2018. He worked on various committees at universities and colleges.

He was a member of the Board of Studies in Electronics at Dr. Babasaheb Ambedkar Marathwada University (Dr. BAMU), Aurangabad. He is a guest editor in STM journals; elected as executive council member (EC) and vice chair of IETE Pune Local Center (2022−24). His research interests are in ad hoc networks, wireless sensor networks, and wireless communications with specific focus on energy and bandwidth.

Prof. Sunil Kr Pandey, D.Sc. (Comp. Sc.), with over 25 Years of experience, presently working with I.T.S, Mohan Nagar, Ghaziabad, India as Professor & Director (IT & UG). He has a strong academic research track record with interest in Cloud, Blockchain, Database Technologies & Soft Computing and is credited 12 Patents granted in India & abroad, 01 Copyright Registered. published 65+ Research papers (SCI/ Scopus Indexed)/ Book Chapters/ Articles with reputed publishers including Springer, IGI, IEEE, Wiley, Hindawi, reputed Journals/ Conferences. He has also published 04 Books on various topics in the field of Soft Computing including Machine Learning, Smart Technologies and IoT with Springer, River Press etc. He is a regular author of Articles in different Print and Online Platforms including Interviews, Views and has published 11 edited volumes on different relevant themes of Information Technologies. He has been providing & coordinating training and consultancy to various reputed organizations including Indian Air Force, Manipal Group, GDA etc. and has been speaking at different forums in India and abroad as speaker, panelist, Expert lectures and guest talks in various Summits/ Conclaves/National/International /conferences/Seminars. He is also a recipient of various awards & recognition from Academia and Industry including Dr. APJ Abdul Kalam Technical University Lucknow, CCS University Meerut, Global CIO Forum, APAC News Media, GEC Media, Business World, Dataquest, Business Standard, CISO Platform, IT Next Magazine, 9.9 Media Group, Enterprise IT Magazine, TechPlus Media Group etc. He has been associated with various Professional organizations including Sr. member of IEEE, ACM, Life Member of CSI, ISCA, IETE etc.

Dr. Albena Mihovska holds a PhD degree in EEng (2008) from Aalborg University, Aalborg, Denmark. She has a strong academic research track record of more than 20 years, with positions as an Associate Professor at Aalborg University, and later at Aarhus University, Denmark. Currently, she holds positions as a CTO, SmartAvatar BV, The Netherlands and as a 6G Research Director at CTIF Global Capsule (CGC) Foundation, Skagen, Denmark, of which she is a Founding Board Member. She is a WG3 Vice-Chair of the One6G

Association (https://one6g.org) and a Board Member of EUROMERSIVE (https://euromersive.eu). Her research interests are in 6G connectivity solutions for applications across vertical market segments. She is an ITU-T Focus Group on Metaverse (FGMV) Expert, and has been active within ITU-T standardisation since 2011. She is an IEEE Member and serves on the Steering and Organizing Committees of several prominent IEEE events.